About this book

Environmental
Pollution
Analysis

P.D.Goulden

In recent years there has been widespread realization of the importance of monitoring the environment to ensure that man's activities do not result in a deterioration in its quality.

The purpose of this monograph is to describe the methods by which environmental samples are obtained and how the levels of particular pollutants in these samples are determined. Throughout this book the emphasis is placed on the most recently developed techniques.

The description of the chemical and instrumental techniques used in the identification and determination of environmental pollutants is sufficiently detailed to enable newcomers to the field to carry out routine analyses. The extensive bibliographies provide the necessary guidance for the extension of background knowledge to any desired depth.

AL POLLUTION
ANALYSIS

HEYDEN INTERNATIONAL TOPICS IN
SCIENCE

Editor: L. C. Thomas

ENVIRONMENTAL POLLUTION ANALYSIS

P. D. GOULDEN

Fisheries and Environment,
Canada

LONDON · PHILADELPHIA · RHEINE

Heyden & Son Ltd., Spectrum House, Hillview Gardens, London NW4 2JQ
Heyden & Son Inc., 247 South 41st Street, Philadelphia, PA 19104, U.S.A.
Heyden & Son GmbH, Münsterstrasse 22, 4440 Rheine/Westf., Germany

ISBN 0 85501 228 5

Printed in Great Britain by Galliard (Printers) Ltd., Great Yarmouth, Norfolk

CONTENTS

FOREWORD

The object of this series of monographs is the timely dissemination of essential information about topics of current interest in science. Interdisciplinary aspects are given fullest attention. The series aims at the presentation of new techniques, ideas and applications in sufficient detail to enable those who are not specialists in a particular subject to appreciate the applicability of the subject matter to their own work, and the bibliographies included in each monograph will guide readers in extending their knowledge of the subject to any desired depth. The depth of treatment, of course, makes them compact definitive books for the specialist as well. The series will from time to time include more general reviews of selected areas of scientific advancement, for which a somewhat wider readership is envisaged.

The topics and the depth of treatment should suit both the student and the research worker, academic or industrial. The range of topics in this series will eventually span the whole extent of scientific interests and the authorship will reflect the international nature of the subject matter.

The widespread realization of the importance of monitoring the environment, to ensure that man's activities do not result in an unacceptable deterioration in its quality, clearly indicated that a book on this subject should be included among the early titles in this series. The present volume is the result: it considers sampling methods for pollutants in air, soil and water; analytical techniques for metals, inorganic non-metals, radionuclides, organic compounds and microorganisms; and concludes with a discussion of continuous monitoring techniques.

The description of the chemical and instrumental techniques used in the identification and determination of environmental pollutants is sufficiently detailed to enable newcomers to the field to carry out routine analyses. The extensive bibliographies provide the necessary guidance for the extension of background knowledge to any desired depth and, as a result, this volume will be of value to all who are interested or involved in the monitoring or control of environmental pollution.

L. C. Thomas

PREFACE

Pollutants in the environment are of concern to people working in a number of diverse fields. Biologists are concerned with the effects on plant and animal life of the materials in the water, air and soil; physicians are concerned with the effect on health of the materials that are in the food we eat, the water we drink and the air we breathe; engineers are concerned with the effect of chemicals and suspended material in the water and air used for domestic and industrial processes. The legislators must decide what levels of pollutants in our environment we are willing to accept in order to enjoy the benefits of modern technology. In these decisions scientists are called upon for advice as to what are 'safe' levels that industries may discharge. Regulatory bodies are concerned with the implementation of the legislation so that these 'safe' levels are not exceeded.

In the early days of environmental concern, analytical methods were not very sensitive and many pollutants could not be detected in air and in natural water a short distance from the emission source. Under these conditions the desirable levels were often defined as 'zero' levels, meaning in fact levels that could not be detected by the available analytical techniques. In recent years great advances have been made in analytical methodology: for example, it is now possible to determine levels of the metallic elements at fractions of a part per 10^9 (parts per billion) and to identify and measure organic materials at levels of one part per 10^{12} (parts per trillion).

The development of these specialized techniques for environmental analysis has tended to make the field an esoteric one. The analyst naturally has an understanding of the methods, their capabilities and their limitations. However, other workers in the field who are concerned with setting up sampling and control programmes, and making decisions from the data produced, do not necessarily have the same detailed knowledge and understanding. As a result the right questions may not be asked at the beginning of a programme, and the data obtained may not be collected in the most efficient way. This monograph has been written with the intention of providing a means whereby non-analytical specialists may become more familiar with how environmental samples are collected and analysed and at what levels particular pollutants may be determined.

Burlington, Ontario   P. D. Goulden
December 1977

INTRODUCTION

A work on Environmental Pollution Analysis should properly begin with a definition of 'pollutant'. This term has been defined as any substance which changes the natural composition of the environment. However, the environment has been changing since the world began, and its natural composition is difficult to identify. Man has played, and continues to play, a major part in bringing about changes and for the purpose of this discussion the term pollutant means a material that enters the environment primarily as a result of man's activities.

Most of the pollutants enter the environment as emissions to the atmosphere or as discharges to water bodies. These may be either in concentrated point sources, such as from factory smoke stacks and sewage discharges, or in a diffuse form such as from automobiles' exhaust and run-off from agricultural land. Most of the material emitted to the atmosphere eventually returns to the earth as particulate fall-out or with rain and snow. Where these materials return over land they may be absorbed by the soil and eventually by the vegetation or they may be washed into the waterways.

The materials discharged to the rivers and streams flow via larger rivers and lakes to the oceans. On the way they may become incorporated into the sediment. Alternatively they may be metabolized by the plant and animal life in the water and thus enter into the food chain. Hence, wherever in the environment the samples are collected, they represent the same process of pollution although occurring in different time spans. Air samples and water discharge samples give a measure of the pollution occurring on a day-to-day, or minute-to-minute, basis. Soil, vegetation, sediment, lake water samples, etc., can give a measure of the pollutants released over a longer time period, often representing in fact an integration of the pollution that has occurred over previous months or years.

In the past it has often been the practice to consider the measurement of air pollution as being a distinct area from, for example, the measurement of water pollution. However, as analytical techniques have become more sophisticated (and analytical equipment has become more expensive), in many cases the final measurement process for a particular pollutant is the same, whether the original sample was of air, water or sediment. The difference in the methods used for the analysis of the various types of sample lies in the sampling technique and in the pretreatment given to the sample to convert it to a form amenable to analysis. Hence the format chosen for this work is to organize the description of the measuring techniques by specific pollutants and, where necessary, to describe how the various types of environmental samples are processed to use this measuring technique.

CHAPTER 1

SAMPLING METHODS

When a decision is made to determine the level of a particular material in part of the environment, the first step is to formulate clearly why the information is required and how the analytical results will be used. Typical objectives may range from the desire for general information on a number of parameters, for the purpose of obtaining baseline data, to a legal requirement to establish the loadings or emissions of a specific pollutant. As will be seen in the later discussion, at many stages of the sampling-analysis process there are decisions to be made as to the specifics of the procedure to be followed. These decisions can best be made if there is a clear understanding of the rationale for the analyses being carried out.

To determine levels in a river, in the air, etc., at a specific location, it is necessary to take samples. The location of the sampling points, the frequency of sampling, the size of the samples and the analytical procedures used are all dependent on the objectives of the analytical programme. The samples should be as representative as possible of the area being sampled. This is important, not only at the time they are taken but also when they have been transported to the laboratory for analysis. Most often this entails that some preservation technique be used. In a programme looking at levels of pollutants in the environment perhaps the most important part is ensuring that valid sample preservation techniques are used and confirming that the analytical results obtained in the laboratory do represent the levels in the samples as taken. Montgomery and Hart[1] have given an excellent discussion of the design of sampling programmes for rivers and effluents which also has application to sampling programmes in any area of the environment. Complete details of the methods by which the various types of environmental samples are taken can be found in the literature references given. Below is given a brief description of these methods for the environmental samples most commonly taken. Also described in general terms are the procedures by which the samples are converted to a form in which they can be analysed.

WATER

For sampling a river or lake from, for example, a bridge, dock or small boat, a bottle of polyethylene or glass of one- or two-litre capacity is used. After determining the depth of the water the bottle is placed in a metal holder attached to a string and thrown into the water. By regulating the rate at which the bottle is lowered to the bottom it is possible to obtain a sample that approximates an integrated sample of the water between the surface and the bottom. However, the bottle should not be allowed to touch the bottom in the sampling to avoid stirring up the sediment. To take a sample at a specific depth the simplest procedure is to immerse the stoppered bottle to the required depth with some arrangement for the removal of the stopper by pulling another string.

There are a number of samplers which were primarily designed for oceanographic sampling, such as the Knudson bottle and the Van Dorn bottle. These are attached to wires hanging in the water and, by a 'messenger' system, allow samples of water to be taken at particular depths. Other similarly suspended samples allow an integrated sample to be collected over a specific range of depths. A description of water samplers is to be found in the ASTM Standards.[2]

Many waters as sampled contain suspended material. If it is desired to include this material in the analysis and to determine the 'total' composition, an appropriate preservative is added to the sample as taken. If only the soluble portion of the water sample is to be determined, the sample is filtered when it is taken and before the preservatives are added. By general definition the 'soluble' portion of a water sample is that which passes through a 0.45 μm membrane filter.

In any natural water sample, chemical and biological processes are occurring when the sample is taken, and these will continue in the sample bottle. In addition, there may be interaction with the bottle, such as adsorption of ions on the bottle wall. To stop, or at least to slow down, these reactions, preservatives are added to the sample. There is no universal preservative available, the most common technique that is effective for many systems being to bring the sample to a pH below 2, and to refrigerate the sample at 4 °C. The low pH stops much of the biological action and retards the adsorption of many cations on the container wall. Some parameters require a special preservative system, for example:

Mercury: per 100 ml sample, 1 ml H_2SO_4 and 1 ml of a 5% solution of $K_2Cr_2O_7$ are added;
Phenol: per 100 ml sample, H_3PO_4 to bring the sample to a pH of 4 or less and 0.1 g of $CuSO_4 \cdot 5H_2O$ are added;
Cyanide: NaOH is added to bring the pH to 12.

For many parameters, such as orthophosphate, sulfite, etc., a satisfactory chemical preservative is not available and the best that can be done is to refrigerate the samples at 4 °C and to analyse them as promptly as possible.

Hence, when sampling, it may be necessary to collect several bottles of water at each sampling point, depending on the parameters to be determined. The specific preservatives required are given in the descriptions of the analytical procedures used. In the case of the determination of microorganisms, it is desired to keep them alive. Hence sodium thiosulfate is added to destroy available chlorine and heavy metals are complexed with EDTA (ethylene diamine tetra-acetic acid).

SEDIMENT

To obtain a sample from the bed of a river, or a lake, a variety of grabs are available. These consist essentially of a bucket having jaws which are either spring-loaded or weight-loaded open. The sampler is lowered to the bottom on a rope and the jaws snap shut to take a sample. The trigger for the jaws to shut may be the weight of the sampler resting on the bottom or a 'messenger' weight dropped down the rope.

There are a variety of designs suitable for use from a small row-boat, such as the 'Ekman' or the 'Ponar' grab. These take samples of about one to two litres in volume and are available from scientific supply houses. To take samples that retain the layer structure of the sediment a core sampler is used. This is a tube which is dropped to the bottom and penetrates to take a cylindrical sample. A valving arrangement at the top of the tube stops the sample from falling out as the sampler is brought to the surface. If the sample is for bacteriological studies a sampler used by Van Donsel and Geldreich[3] is recommended because of the difficulties in maintaining aseptic conditions in the field with other samplers.

There are a number of grab samplers designed to be used from a large vessel, such as the Franklin–Anderson, the Peterson and the Shipek samplers. These take a much larger sample from the lake bed and with these the top ten centimetres or so of the structure is retained. Samplers used for sampling sediment in the Great Lakes have been discussed by Sly.[4] No chemical preservatives are added to the sediment sample; immediately it is taken it is placed in a plastic bag and refrigerated or stored in an ice-chest. If the sample is not to be analysed on the day it is taken, it is immediately deep-frozen.

To carry out an analysis the sample, as taken, or the thawed-out sample if it has been frozen to preserve it, is passed through a 10-mesh screen to remove large pieces of foreign matter such as twigs or stones. If necessary the sample can be pressed through the screen with a rubber stopper. The sample is then treated in a blender (such as a 'Waring' blender) set at high speed, to homogenize it. Portions of this homogenized sample are taken for the various analyses desired.

Since the most unambiguous basis for reporting results is on a dry basis, a portion of the wet homogenized sample is oven dried at 105 °C. The loss in weight in this drying provides a conversion factor by which the analysis carried out on the other portions of the sample can be converted to a dry-weight basis.

In the determination of the inorganic constituents of the sediment there are two ways to look upon the analysis. If the concern is to determine the materials which are adsorbed on the mineral particles, then the analysis is carried out upon a solution made by extracting the sediment with, for example, an acid. This does not dissolve the mineral particles but is believed to solubilize the metal ions that are adsorbed upon them. The rationale for this approach is that the concern is for those pollutants which have been deposited in the sediment and which might be available for biological processes at the sediment–water interface, rather than for the geochemical composition of the minerals on the lake bed.

Hence a variety of dilute mineral acids are used to extract the anions such as nitrate, sulfate, and phosphate which are adsorbed on the sediment. A mixture of concentrated nitric acid and hydrogen peroxide is used to extract the metals. Materials which are volatile, such as ammonia, phenol, and cyanide (as HCN), are separated from the sediment by distillation. To determine organic materials, herbicides, pesticides, etc., the sediment is extracted with organic solvents.

If the concern is for a 'total' analysis, i.e. the chemical constitution of the total sediment, it is necessary either to employ an analytical technique that uses solid samples, or to solubilize all the minerals completely. Classically this solubilization has been done by a fusion with a variety of inorganic salts, but the most convenient way now is to use a 'Parr' bomb. This is a 'Teflon'® capsule enclosed in a steel shell which allows solubilization of the sediment with hydrogen fluoride under pressure. The technique was described by Bernas[5] and has been used for a variety of environmental samples besides sediments.

FISH

Fish when caught are classified by species, size and estimated age. The fish are cooled in ice or refrigerated as soon as possible after they are caught. If the whole fish is to be analysed it may be preserved by being deep-frozen immediately. If certain parts of the fish, such as the liver or muscle tissue, are to be analysed, the separated organs can be preserved by being deep-frozen.

The fish, or part of the fish, is homogenized in a blender. Portions of this homogenate are then weighed for the analysis. For the determination of organic pollutants the fish tissues are ground and extracted with organic solvents. The treatment and clean-up procedures used on these extracts are discussed in the chapter on the analysis of organic pollutants (Chapter 6).

To analyse for inorganic material by methods using a sample in solution, the sample is ashed and the inorganic residue solubilized. There is a choice in ashing procedure between dry-ashing and wet-ashing. In dry-ashing, the sample is heated to 400–600 °C in an open dish and oxidized by the ambient air. In wet-ashing the sample is treated with a variety of oxidizing systems such as nitric–sulfuric acid, perchloric–nitric–sulfuric acid, sulfuric acid–potassium permanganate, and hydrogen peroxide.[6] Dry-ashing has the advantage that the minimum amount of chemicals are added and hence the reagent blank for trace metals is reduced. However, some of the metals may combine with the vessel in which ashing is carried out and volatile elements such as arsenic, selenium and mercury may be lost.

Rather than combustion in an open dish, the sample may be burnt in an oxygen combustion flask. The sample is dehydrated by being stored over phosphorous pentoxide and then ignited in a flask filled with oxygen. Since it is a closed system the volatile materials are retained. Procedures for the determination of selenium[7] and mercury[8] using this technique have been described.

Another option for dry-ashing is plasma-oxidation where the organic sample is treated with an oxygen-bearing plasma.[9] In this case the oxidation proceeds at temperatures of around 100 °C so that the loss of volatile materials is minimized.

A discussion of ashing procedures is given in the texts on atomic absorption spectroscopy (see Chapter 2). The preparation of biological samples for multi-element analysis is discussed by Hamilton et al.[10]

AIR

There are three methods of sampling air: a static sensor; taking a whole-air sample; and separating the pollutants at the sampling site.

A static sensor is the simplest form of air sampler. This uses the natural movement of air past it to present the sample. One example of this is the dustfall jar, which consists of a container that is used to collect particulate matter that falls freely into it. Another example is the 'candle' or plate used to measure the sulfur-containing gases. These consist of a matrix containing lead peroxide which combines with these gases to form lead sulfate. After being exposed to the air for a time the candle is taken to the laboratory, the matrix is dissolved and the amount of sulfate produced is measured. Another type of static sensor is the CDE Toxic Hazard Monitor.[11] This consists of a small holder containing an adsorber, such as a piece of charcoal cloth, covered with a semipermeable membrane. The pollutant vapour diffuses through the membrane, is adsorbed by the charcoal and is later eluted for measurement. The sensor may be used for atmospheric sampling or, attached to clothing, it may be used as a personal exposure monitor by people working in a contaminated work area. Determinations of air pollution may also be made by exposing indicator paper to the air and comparing the colour produced after a certain time with a standard chart.

For a whole-air sample, air is taken in a container for later analysis in the laboratory. This air may be collected in a previously evacuated vessel by merely opening a valving device to the atmosphere being sampled. In the flasks described in ASTM Standards[2] a glass flask (of up to about 500 ml volume) is evacuated and the end of the connecting tube is sealed with a flame. To take the sample the end of the tube is broken to admit the air sample; after sampling, the end is closed with a rubber cap or wax-filled cartridge. Alternatively the sample may be drawn into a container by a rubber aspirator bulb, a vacuum pump, or by water displacement. With the aspirator bottles samples of up to five litres are normally collected. Plastic bags can also be used to collect the sample; the bag is purged and filled with a rubber squeeze bulb or other pumping device.

Whole-air sampling has the advantage that very little equipment is needed in the field in order to take the sample. It is often a convenient way to take a sample for an organic pollutant where there is no well-established adsorbent available and where the sensitive detection techniques of gas chromatography can be used to overcome the limitation of the sample size. There are possible problems of sample preservation since reactive gases such as hydrogen sulfide, oxides of nitrogen, sulfur dioxide, etc., will react with dust particles, moisture and perhaps the container material itself.

Separation of the pollutants at the sampling site is the most usual way of sampling. The pollutants are separated from a known volume of air and analysed later in the laboratory (or in the field). For the collection of particulate matter a number of sampler types are available. The particulates may be separated by drawing air at a known rate through a filter medium. They may also be collected with an impingement device. The air is blown at a surface and because of their inertia the particles collect on this surface. A microscope slide as the collector surface is often used. This facilitates later examination of the particles. The surface may be dry, in which case the apparatus may be called an impactor, or the impingement surface may be covered by a liquid. The particulates may also be collected by electrostatic precipitation or by thermal precipitation.

The particulates may be collected for three reasons: to determine their total amount, to determine their particle size distribution and/or to determine their chemical composition. The particle size distribution may be determined by physical examination of the collected particulates, or the different sizes may be separated by the collector itself, e.g. a series of impactors in cascade. To collect particulates for chemical analysis, typically a filter in a high-volume sampler is used.

For elemental analysis where a solution is required, such as the determination of the metal content by atomic absorption, the particulates are dissolved in acid. A more complete solution can be made by digesting with hydrogen fluoride in a 'Parr' bomb. This also has the advantage that refractory materials such as silica in the particulates can be determined.[12] Organic materials are recovered by extraction with solvents. For analysis where solid samples can be used, such as in X-ray fluorescence, pieces of the filter are used directly.

Collection of gases and vapours is usually by absorption in a liquid. The air sampled is first cleaned from particulates by passing it through a filter. This air is then passed through an absorber containing a liquid which extracts the pollutant of interest. Various absorbers are described in the ASTM Standards.[2]

For gases which are readily soluble distilled water may be used in the absorber. Often a reagent which will react chemically with the pollutant is used because this solution may be used later in the analytical procedure.

Adsorption of the gases and vapours on to solid materials is another method of collection. Such adsorbents as activated carbon, silica gel, or activated alumina are used. Sometimes the adsorbent is made more specific for a pollutant by impregnating the solid with an appropriate chemical. After exposure to the air stream the pollutant is recovered from the adsorbent by heat or by chemical treatment. A discussion of adsorbents and techniques is given in the ASTM Standards.[2] Separation of gases and vapours may be made by condensation where the gas is passed through a trap at low temperature. This may be combined with adsorption to separate a whole range of pollutants.

For measurements in the field, indicator tubes can be used. The air is drawn through a tube containing reagents that react with the pollutant to form a colour, the amount of colour produced being compared with colour standards. Indicator tubes are available for over a hundred pollutants; these are listed by Saltzmann.[13]

Many of the 'standard' pollutants such as nitrogen oxides, sulfur dioxide, ozone, etc., which in the past were measured by absorption into a reagent solution and determination of the resulting colour, can now be more conveniently measured by direct reading instruments. Particularly, the application of chemiluminescent methods[14] to the measurement of air pollution is resulting in the use of the direct reading instruments for some pollutants being preferable to the taking of an air sample and analysis by wet chemistry. These methods are discussed in the chapter on continuous monitoring.

A comprehensive list and description of air-sampling instruments is given in a publication by the American Conference of Governmental Industrial Hygienists.[15]

VEGETATION

When sampling vegetation such as leaves from trees, it is important that the samples be carefully identified with respect to the species of the plant, the age or maturity of the foliage and the age of the plant. This ensures that the data obtained can be related to data from other sampling sites. Samples of foliage from trees are taken by cutting the outside growth from ground level up to about 20 ft or more in height and collecting all the leaves. A useful sample size is 500–1000 g. If a particular source of air pollution is being investigated, it is preferable to sample the sides of trees that face this source. When sampling field crops or grass, leaves or blades are cut at regular intervals in a network across the field. Flower heads and stalks are not included in the sample, nor is root material. Any areas of very local contamination such as those receiving roadside dust should preferably not be sampled, unless of course it is this type of contamination that is being studied.

The samples are collected in perforated polyethylene bags and placed in refrigerated storage or an ice-chest as soon as possible. They can be safely stored in a refrigerator for a few days.

Collecting vegetation samples may be for two reasons: to determine the amount of contamination on the surface of the leaves or to determine the chemical composition of the leaves themselves. When the samples are received in the laboratory they are divided into two parts. The vegetation in one of these parts is carefully sponged with water containing a detergent and then well rinsed with de-ionized water to remove all the surface contamination. Analysis of each of these parts then gives the composition of the leaves and the composition of the leaves plus surface material. From these the amount of surface contamination can be calculated. For determination of organic materials the sample is ground and extracted with a variety of organic solvents. For inorganic analysis the vegetation is dried in a forced draught oven at 80 °C for 24 h. The dried sample is then crushed and homogenized, and stored in a jar; the sample in this state is stable indefinitely. For analysis by methods using a solution as the sample, the dried vegetation is ashed, the considerations of the method used being the same as those discussed above (pp. 4–5).

SOIL

Soil is sampled with a core sampler, which is a stainless steel tube about 2 cm in diameter. This is pushed into the soil to take a sample about 15 cm in depth. On removal the core is separated into three portions of about 5 cm each, which are treated as separate samples. A number of cores are taken at each sampling point. Ten is a normal number, the ten 5-cm portions all being placed in a plastic bag as one sample. The samples should be refrigerated as soon as possible and if they are not to be analysed immediately they are preserved by being deep-frozen.

The considerations of the analytical techniques to be used are the same as those discussed above under 'Vegetation'.

PRECIPITATION

The simplest type of rain sampler is essentially a funnel and bottle. The rain and snow fall into the funnel and hence into the bottle. The narrow neck of the bottle minimizes the loss of water by evaporation. This type of sampler also collects the dust that falls when it is not raining and hence the sample represents the total fall-out of pollutants.

To collect only precipitation samples an automatic sampler is used which has a sensor that detects when rain falls upon it. When it is not raining the collection vessel is closed; when the sensor sees rain or snow the lid of the vessel is opened and the sample is collected.

REFERENCES

1. H. A. C. Montgomery and I. C. Hart, *Water Pollut. Control* **73**, 77 (1974).
2. ASTM Standards, 1976 Annual Book of ASTM Standards, Part 26, American Society for Testing and Materials, Philadelphia, Pa., 1976.

3. D. J. Van Donsel and E. E. Geldreich, *Water Res.* **5**, 1079 (1971).
4. P. G. Sly, *Proc. 12th Conf. Great Lakes Res.* 883, 1969.
5. B. Bernas, *Anal. Chem.* **40**, 1682 (1968).
6. Analytical Methods Committee, *Analyst* **101**, 62 (1976).
7. M. R. Church and W. H. Robinson, *Int. J. Environ. Anal. Chem.* **3**, 323 (1974).
8. I. Okuno, R. A. Wilson and R. E. White, *J. Assoc. Off. Anal. Chem.* **55**, 96 (1972).
9. C. E. Gleit, *Anal. Chem.* **37**, 314 (1965).
10. E. I. Hamilton, M. J. Minski and J. J. Cleary, *Sci. Total Environ.* **1**, 1 (1972).
11. Chemical Defence Establishment, Porton Down, Salisbury, Wilts., England.
12. L. E. Ranweiler and J. L. Moyers, *Environ. Sci. Technol.* **8**, 152 (1974).
13. B. E. Saltzmann, Section S-1, *Air Sampling Instruments for Evaluation of Atmospheric Contaminants*, 4th Edn, American Conference of Governmental Industrial Hygienists, Cincinnati, Ohio, 1972.
14. J. A. Hodgeson, *Toxicol. Environ. Chem. Rev.* **2**, 81 (1974).
15. *Air Sampling Instruments for Evaluation of Atmospheric Contaminants*, 4th Edn, American Conference of Governmental Industrial Hygienists, Cincinnati, Ohio, 1972.

CHAPTER 2

THE DETERMINATION
OF METALS

The term 'metal' can be applied to more than 80% of the elements. The elements that will be addressed here are those that have some metallic characteristics and which are known to be of concern in pollution analysis. These are, first, those which occur at trace levels in most environmental samples, namely Ag, Al, As, B, Cd, Co, Cr, Cu, Fe, Hg, Mn, Mo, Ni, Pb, Sb, Se, Sn, V and Zn. The others discussed are those which occur at high levels in a large number of samples, namely Ca, Mg and Na.

The methods for determining these elements are described in general terms. These methods have in common that for the most part no information is obtained as to the chemical form of the element in the sample. From this aspect this chapter might also be described as dealing with 'elemental analysis'. At the end of the chapter are listed recent papers describing methods of analysis for these elements in a variety of environmental samples.

The methods that will be described here are:

Atomic spectroscopy
Molecular absorption spectrophotometry
Polarography and anodic stripping voltammetry
Neutron activation
X-ray fluorescence

In the first three of these the samples are mostly processed as solutions and for those samples which are taken as solids a digestion procedure as described in Chapter 1 is used. In the last two techniques the samples are processed as solids and hence these methods may be particularly suitable for some of the solid environmental samples. Although included in this chapter in the determination of metals, neutron activation and X-ray fluorescence are means of determining elemental composition and will also measure concentrations of elements other than metals.

Any description, such as this, of instrumental techniques suffers from the fact that the field is in a state of change. In particular, continuing improvement in solid-state electronics is resulting in such changes as microprocessors being used in almost every type of instrument for control and data handling. The performance of these instruments is quite different from that of those used to date. Similarly the cost reductions that have occurred in areas such as multi-channel analysers now make non-dispersive energy measurements much more feasible than they have been before. Hence any comments made as to the relative merits and the extent of use of the different techniques may be obsolete by the time that they are read.

ATOMIC SPECTROSCOPY

There are three fields of atomic spectroscopy that will be considered: atomic absorption, atomic emission and atomic fluorescence. These all operate by measuring energy changes that take place in the atomic state. In all of them the sample is treated to vaporize it and then to cause it to dissociate into its elements in a gaseous state. A flame can be used to do this. In atomic absorption, the atoms in the ground state will absorb energy to go to a higher energy state. Energy as radiant energy of the characteristic wavelength to bring about this transition is shone through the cloud of atoms and the amount of energy absorbed is determined. In atomic emission, the atoms absorb energy from the flame or other high-energy source to get into an excited state, and then emit energy of a characteristic wavelength as they return to a less excited state. In atomic fluorescence, the atoms are excited by being irradiated with energy from an external source, and the light emitted as they fall back to a lower energy state is measured.

Atomic absorption is currently the most widely used technique. It suffers from the disadvantage that it is essentially a single-element determination, but the analysis can be carried out swiftly, the procedures are well understood and there are many excellent instruments available. Atomic emission using a flame is used for a few elements such as the alkali metals and alkaline earths. It is not very sensitive for other elements because of the relatively low temperature of available flames. However, the recent adaptation of plasma technology, particularly the inductively coupled argon plasma (ICAP), where temperatures approaching 10 000 K are obtained, has made atomic emission a practical route to the determination of most metals. It is a technique for simultaneous multi-element determination and hence offers an economic route to amassing a great deal of data. Atomic fluorescence, for many elements, is a more sensitive technique than atomic absorption and it also has the advantage of being potentially a multi-element technique. However, despite its apparent advantages it is not widely used in routine analysis and there are no commercial instruments available. It seems to have been a 'bridesmaid' for many years and in the author's opinion

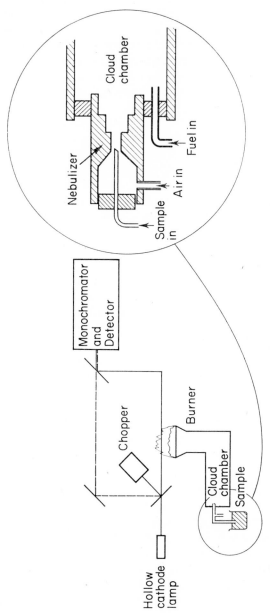

Fig. 2.1. Schematic of atomic absorption.

it has now probably been pre-empted in multi-element analysis by atomic emission using ICAP.

Atomic absorption spectroscopy

In its common form atomic absorption spectroscopy functions as shown in Fig. 2.1. Air is blown through a venturi, and the sample solution is drawn into the throat of the venturi and into the air stream as a spray of fine droplets. A large fraction ($\sim 90\%$) of the drops fall out of the air stream in the cloud chamber but the very small ones remain suspended. The air is mixed with a fuel gas and the mixture passes through a narrow slot where it is ignited to form a flame. In the flame the sample vaporizes and forms a cloud of atoms in the gaseous state. Through the flame is passed a light beam that contains light of the wavelength corresponding to the energy required to raise the atoms of the particular element being analysed from their ground state to an excited state. This wavelength is observed by the monochromator and the amount of energy absorbed by the flame is determined, hence the number of atoms in the ground state in the flame. A hollow cathode lamp may be used to obtain the light of the required wavelength. This is a lamp with a cathode that consists of a hollow chamber containing the element being determined, so that the emission radiation of the element is obtained. Electrodeless discharge lamps are also used for some elements.

The oxidant–fuel mixture most commonly used is air–acetylene. Other mixtures can be used, and these are shown in Table 2.1 with the approximate temperature for each.[1a]

TABLE 2.1
Flames used in atomic absorption

Oxidant	Fuel	Flame temperature/°C
Entrained air	Argon–hydrogen	300–800
Air	Propane	1900
Air	Hydrogen	2000
Air	Acetylene	2300
Nitrous oxide	Acetylene	2600

There are a number of potential interferences in this technique; they may be classified in three categories: 'ionization', 'chemical' and 'matrix'. In ionization interference the element absorbs energy from the flame and becomes ionized. The ions of the element absorb energy at a different wavelength from that of the ground state so that the monochromator and photomultiplier do not 'see' this absorption of energy. In addition the ion, in falling back to the ground state, may emit energy of the same wavelength as the analytical line whose absorption is being measured. Ionization occurs most easily in the alkali or alkaline earth elements. It is a function of the flame temperature, and lowering the temperature will reduce ionization. It is often preferred to determine sodium with the air–hydrogen flame rather than with the slightly hotter air–acetylene flame in order to reduce the ionization interference. However, lowering the flame temperature increases the probability of chemical interferences. Ionization can also be overcome by the addition of a large amount of an easily ionizable element such as one of the alkali metals. The ionization effect in the determination of sodium can be minimized by the addition of, for example, potassium.

Chemical interferences arise when the fraction of the element that forms atoms in the ground state in the flame is different with the sample than with the standard solutions. This may result in a depression or an enhancement of the signal. Where the element forms a compound that is not dissociated readily in the flame the signal is depressed. Such a situation arises, for example, in the case of calcium when phosphate is present. The calcium phosphate is not as readily dissociated in the air–acetylene flame as are the other calcium salts. This type of interference may be overcome by chemical treatment. Lanthanum may be added which forms a stronger complex than calcium and prevents the formation of calcium phosphate. A higher flame temperature reduces this type of chemical interference, and if a nitrous oxide–acetylene flame is used for calcium, phosphate no longer interferes. However, in this case the flame is hot enough to cause ionization interference, and in determining calcium with the nitrous oxide–acetylene flame it is necessary to add an alkali metal to suppress ionization. This flame is used for determining those elements such as vanadium and aluminium which form refractory oxides.

Enhancement of the signal occurs when the element is present in a form that is more readily dissociated than the standard being used. This situation may arise when there is a large amount of organic material in the sample, in a similar way to the enhancement of the signal when chelation-solvent extraction is used (see below). It may be possible to overcome this type of interference by more careful matching of the composition of the standards to the sample, or it may be necessary to convert the sample to a known form, either by digestion or by solvent extraction.

Matrix effects are due to chemical and physical differences between the sample and the standard. If the sample has a high solids content then there may be scatter in the flame from the large number of solid particles formed. This interference is best overcome by the use of a background corrector. In this technique the apparent absorption of a wavelength close to the analytical wavelength is measured. This gives a measure of the non-specific absorption in the flame and can be used to correct the apparent absorption at the analytical wavelength. Many instruments are equipped to carry out this background correction on an automatic basis.

With some samples, the amount of the sample reaching the flame may not be the same as the amount of the standard solutions used to calibrate the determination. This may be because the amount drawn into the venturi is not the same, or the fraction of the droplets remaining in the air stream is different.

Either effect can be due to differences in viscosity, density or surface tension between the standard solution and the sample. This can be overcome by better matching of the properties of standards and sample. However, where the physical properties of the sample are different from those of the standard solutions and it is not practicable to duplicate the properties of the sample in a standard, the technique of standard additions can be used. Known additions of small volumes of a concentrated standard are made sequentially to the sample, and after each addition the absorbance in the flame is determined. Plotting the incremental additions against the absorbance yields a line, which when projected back to the original sample absorbance, gives a measure of the original sample concentration.

The combinations of materials that can cause interferences in the determination of a particular element are numerous. However, one advantage that atomic absorption has is that, in the relatively short period that it has been used, an enormous amount of work has been done in enumerating interference problems in many matrices. This has resulted in the availability of a 'cook-book' methodology, i.e. to determine an element in almost any matrix, a procedure is available which gives reasonable assurance that, if followed, it will yield valid results.

TABLE 2.2
Detection limits for some elements using direct aspiration and a flame

Element	Wavelength/nm	Flame[a]	Detection limit/μg l^{-1}
Ag	328.1	1	2
Al	309.3	2	20
As	193.7	4	50
B	249.7	2	1500
Ca	422.7	1	0.5
Cd	228.8	1	2
Co	240.7	2	10
Cr	357.9	1	3
Cu	324.7	1	1
Fe	248.3	1	5
Hg	253.6	3	250
Mg	285.2	1	0.1
Mn	279.5	1	2
Mo	313.3	2	20
Na	589.0	1	0.2
Ni	232.0	1	2
Pb	283.3	1	10
Sb	217.6	1	40
Se	196.0	4	50
Sn	224.6	3	10
V	318.4	2	40
Zn	213.9	1	1

[a] 1, Air–acetylene; 2, Nitrous oxide–acetylene; 3, Air–hydrogen; 4, Argon–hydrogen-entrained air.

In the form described, where samples are aspirated directly to a flame, atomic absorption will determine most metals in the mg l^{-1} range.

In Table 2.2 are given the conditions used in the determination of the elements most often of concern in environmental analysis. The detection limits for the elements are also given. The detection limit is the concentration that gives a signal that is twice that of the background noise. (In atomic absorption work the sensitivity of the method is often quoted and this is the concentration that gives 1% absorption.) In most analytical work it is desirable to work at levels that are more than five times the detection limit in order to obtain data that is reasonably precise. However, when carrying out analysis on environmental samples, it is often found that a large fraction of the samples contain levels close to the detection limit. In the author's opinion, levels that are found to be greater than twice the detection limit should be accepted as valid data, remembering however that at these levels the relative standard deviation is of the order of 20%.

The limits of detection shown in Table 2.2 approximate to those obtainable in the analysis of water samples, providing care is taken in sample preservation and in avoiding contamination via the sample bottle. For air particulates and biological samples, where a digestion procedure is used, the detection limits obviously depend on the 'ashing' procedure, the sample size and the dilution made after digestion. In the case of air particulates the background materials in the filter are also of concern. Ranweiler and Moyers[2] have described a procedure that measures 22 elements in air particulates by atomic absorption. They quote practical detection limits for these elements; these are shown in Table 2.3.

TABLE 2.3
Practical detection limits for the atomic absorption analysis of atmospheric particulates (from Ranweiler and Moyers, Ref. 2)

Element	$\mu g\ m^{-3}$ in air sample	Element	$\mu g\ m^{-3}$ in air sample
Si	0.8	Sr	0.003
Al	0.4	Ni	0.002
Ca	0.03	V	0.006
Fe	0.07	Mn	0.0003
K	0.04	Cr	0.0003
Na	0.03	Rb	0.0005
Mg	0.01	Li	0.0003
Pb	0.007	Bi	0.002
Cu	0.003	Co	0.0008
Ti	0.07	Cs	0.00004
Zn	0.03	Be	0.00004
Cd	0.0003		

For the large number of metals that occur at the trace level in environmental samples direct aspiration to the flame is not sensitive enough. One technique used to increase the sensitivity is to complex the metal with an organic complexing agent and, after adjusting the pH, to extract the metal complex into an organic solvent. The bulk of the material in the sample, such as the alkali and alkaline earth salts, are left behind in the aqueous phase. This has been discussed by Mulford.[3] The improvement in sensitivity is obtained by three mechanisms. The first is that the volume of the solvent can be small relative to the sample so that a concentration factor of up to 50 may be obtained. The second is that there is enhancement of the signal when an organic solvent is aspirated; this depends on the solvent and metal being determined, but is of the order of a factor of three. The third is that because the extraction system is selective and the large part of the dissolved solids are left in the aqueous portion, there is less 'noise' in the flame. In analysing biological and geological samples, solvent extraction is often used to separate the elements of interest from the matrix even when, on the face of it, a concentration is not necessary. The complexing agents and solvents recommended for a variety of metals are shown in Table 2.4. The pH at which the extraction is carried out is important; the optimum range is also shown in Table 2.4.

TABLE 2.4
Chelate-solvent extraction systems

Element	Chelate	Solvent	pH of Extraction
Ag	Dithizone	Ethyl propionate	3.5–6.5
Al	8-Hydroxyquinoline	MIBK[a]	8.0
Cd	APDC[b]	MIBK	1.0–5.0
Co	APDC	MIBK	2.0–4.0
Cr	APDC	MIBK	3.5 ± 0.2
Cu	APDC	MIBK	1.0–3.5
Fe	APDC	MIBK	1.5
Mn	8-Hydroxyquinoline	MIBK	9.0–11.0
Mo	Oxine	MIBK	2.0–2.4
Ni	APDC	MIBK	2.0–4.0
Pb	APDC	MIBK	2.8 ± 0.2
Sb	APDC	MIBK	3.5 ± 0.2
V	8-Hydroxyquinoline	MIBK	3.0 ± 0.2
Zn	APDC	MIBK	2.5–5.0

[a] MIBK—methylisobutylketone; [b] APDC—ammonium pyrrolidine dithiocarbamate.

A typical procedure for the determination of a metal by chelation-solvent extraction is that used, for example, in the determination of lead in a water sample. In this procedure a 150-ml sample is placed in a 250-ml volumetric flask. Sodium acetate buffer (8 ml) and 8 ml of a 1% ammonium pyrrolidine dithiocarbamate (APDC) solution are added. (The buffer is calculated to bring the pH of the solution to 2.8, assuming that the sample has been preserved by the addition of 0.2% HNO_3.) Methylisobutyl ketone (MIBK) (10 ml) is added and the flask is shaken vigorously and then allowed to stand for a few minutes for the layers to separate. A 'floating' solution is added slowly down the side of the flask to bring the organic solvent layer up into the neck of the flask. The floating solution is a solution of buffer at the same pH and approximate ionic strengthas the buffered sample. It is saturated with solvent by having been previously shaken with MIBK. The organic layer in the neck of the flask is aspirated into the atomic absorption burner.

It is difficult to define the detection limits obtained by chelation-solvent extraction, since the aqueous:solvent ratio can be increased in special circumstances to improve the sensitivity. The limit then often becomes the background error introduced in the sampling-processing operation rather than the actual measurement itself. It is, however, possible to measure the metals shown in Table 2.4 at levels of a fraction of $\mu g\ l^{-1}$ in water samples, using the systems shown.

When it is desired to analyse several elements in one sample the solvent extraction may be adapted so that with one extraction all the elements of interest can be extracted and then determined sequentially on the one extract. It is seen from Table 2.4 that Cd, Co, Cr, Cu, Pb, Ni and Zn can all be extracted with APDC and MIBK at a pH of 3. Kinrade and Van Loon[4] describe a procedure for extracting eight metals (Cd, Co, Cu, Fe, Pb, Ni, Ag and Zn) from natural waters with MIBK using a mixture of APDC and DDDC (diethylammonium diethyl dithiocarbonate).

Another way to obtain a concentration-separation, particularly applicable to water samples, is to use a chelating resin to selectively extract the heavy metals of interest. Dingman et al.[5] have reported a concentration factor as high as 1000 when using polyamine–polyurea resins to extract Cu, Zn, Ni and Co. The metals are eluted from the resin with acid and determined by atomic absorption.

An alternative approach to achieving sensitivity in atomic absorption is the use of the 'flameless' techniques. In this process a graphite tube is placed co-axially in the light path of the instrument. Provision is made to heat the tube electrically to close to 3000 °C. The tube is blanketed in an inert gas, such as nitrogen or argon. Using a micropipette a sample (5–100 μl) is injected into the graphite tube. The tube is heated to dry the sample, then the temperature is raised to about 1000 °C to pyrolize organic matter. The tube is finally heated to a temperaure of about 2700 °C to atomize the sample. The absorption during this atomization period is measured.

By using a small graphite cuvette or micro-boat to introduce the sample, the drying stage can be carried on outside the tube, and this also gives the option of handling solid samples. However, the amount of solid that can be used is small, of the order of 50 mg maximum, and at this level homogeneity of the sample becomes a consideration.

Other variations of the same technique use an electrically heated tantalum ribbon rather than a graphite tube as the means of atomizing the sample.

The absolute detection limits with the flameless technique are very low; typical values are shown in Table 2.5. The detection limits in terms of concentration can be calculated from these, assuming that 100 μl of sample are used

TABLE 2.5
Detection limits by various atomic spectroscopy techniques

	Flameless A.A. (Abs. det. limit) pg.	Flame emission $\mu g\,l^{-1}$	Atomic fluorescence $\mu g\,l^{-1}$	ICAP[a] $\mu g\,l^{-1}$
Ag	0.5	8	0.1	4
Al	5	5	5	2
As	10	—	100	40
B	—	—	—	5
Ca	5	0.1	0.001	0.07
Cd	0.5	800	0.01	2
Co	50	30	5	3
Cr	10	4	4	1
Cu	5	10	1	1
Fe	5	30	8	5
Hg	200	—	20	200
Mg	0.5	5	0.1	0.7
Mn	1	5	2	0.7
Mo	50	100	60	5
Na	—	0.1	—	0.2
Ni	100	20	3	5
Pb	5	100	10	8
Sb	20	600	50	200
Se	50	—	40	30
Sn	100	300	50	300
V	500	10	70	6
Zn	0.1	—	2	2

[a] From Ref. 9—Ames Laboratory Data.

The sensitivity can be further increased, and matrix effects minimized, by a chelation-solvent extraction. For flameless work the preferred solvent is often chloroform. The APDC complexes are as soluble in this as in MIBK. Methyl-isobutyl ketone is used in the flame because it has good burning characteristics.

Mercury is an element whose level in the environment is currently of considerable concern. The conventional atomic absorption techniques are not sufficiently sensitive to measure the levels found in environmental samples. However, mercury vapour at ambient temperatures is in the atomic state and this enables mercury to be determined by the 'cold-vapour' technique. In this, the mercury in the sample solution is reduced to the elemental state and then swept from the solution in a stream of air. This air is passed through a quartz-windowed absorption cell in the light path of the atomic absorption spectrometer and the absorption at 253.7 nm is measured. This is a very sensitive method and enables levels down to 0.01 $\mu g\,l^{-1}$ Hg in water samples to be determined. The reduction is usually carried out using a stannous salt in an acid solution containing chloride ion and hydroxylamine. In this system only inorganic mercury is reduced, so that any organomercurials must first be oxidized. This is usually done by a digestion with potassium peroxydisulfate and potassium permanganate. The process can be carried out on a manual basis or in an automated Auto Analyzer® system.[6] Rather than using a spectrometer the absorption measurement may be measured simply by a mercury lamp and a photometer. A possible interference in this technique is absorption by volatile organic materials. However, the digestion used to oxidize the organomercurials should also destroy the organic materials and if a spectrometer is used for the absorption measurement this destruction of organics can be verified by means of the background corrector.

Other elements that are difficult to measure at environmental levels by the conventional techniques are arsenic and selenium. This is partly because their absorbing lines are at low wavelengths (<200 nm) and at these wavelengths the acetylene flames themselves give considerable absorbance and noise. The argon–hydrogen-entrained air flame is more transparent at these wavelengths but its low temperature gives opportunity for chemical interferences. In these cases, use is made of the fact that the elements can be reduced to form volatile hydrides, and hence can be separated from the sample matrix. The hydrides are then decomposed in a low-temperature flame. The hydride generation not only gives separation from the matrix but also an opportunity to concentrate the arsenic and selenium. The most common method of generating the hydride is reduction with sodium borohydride in acid solution, Other metals that can be determined in this way are antimony, bismuth, tin and lead. An alternative to using a conventional burner for the production of the atomic elements from the hydrides is to carry out the combustion in an open ended tube in the light path of the atomic absorption (AA) spectrometer. This gives an increase in sensitivity of several orders of magnitude over the use of a conventional burner. This hydride generation technique can be used either on a manual basis or in automated equipment.[7]

Atomic absorption spectroscopy has proved to be a versatile tool in measuring concentrations of trace metals. The one big drawback to its use is the fact that typically only one element at a time is determined. However, there are a few instruments made which have two monochromators and allow two simultaneous measurements. Other approaches to multi-channel operations are the use of multi-element hollow cathode lamps with vidicon detectors and echelle diffraction gratings.

Atomic emission

Atomic emission spectroscopy determines the concentration of an element by measuring the amount of radiation emitted as an excited atom goes to a less excited state. The attraction of atomic emission techniques is that they provide a means for simultaneous multi-element analysis. Since all the atoms present in the sample may be excited by a common source the resolution of the line spectrum emitted enables the concentrations of all the species present to be determined. The various forms of atomic emission differ in the means by which the excitation is carried out. Arc-spark discharges have been used for a long time as excitation sources for the determination in environmental samples of elements at other-than-trace levels. For the easily excited elements such as the alkali metals a flame will excite a sufficient fraction of the atoms present to enable a sensitive analysis to be performed. However, for a large number of metals even the very hot nitrous oxide–acetylene flame does not enable a flame-emission analysis to be practical. The detection limits obtained by flame-emission are shown in Table 2.5.

An excitation source that has received a great deal of attention, particularly recently, is a plasma. A plasma is a gas in which a significant fraction of the atoms or molecules are ionized. This means that electrical energy can be transferred to it. If it is sufficiently ionized a large amount of energy can be supplied to it and heat it to a very high temperature. In flames the temperatures obtained are limited by the heats of combustion; there is no such limit in plasma sources. In addition, since a plasma can be generated in an inert atmosphere, there are less chemical interferences than in the highly reactive flames. There are a variety of ways of generating plasmas, and various plasma emission sources used in analytical spectroscopy have been reviewed by Greenfield et al.[8] They also discuss the concept of temperature in high temperature excitation sources.

The system that will be described here is the inductively coupled argon plasma (ICAP) since this appears at the moment to be the system which will be most widely used. This has been described by Fassel and Kniseley[9] and there are many articles in the literature regarding its application. The power input to the plasma is by a radio frequency generator operating in the 4–50 MHz range with a power output of up to 5 kW. The current from the generator is fed to a coil placed around a quartz tube through which flows argon. The oscillatory current flowing in the coil produces an oscillating magnetic field with the lines of force aligned axially along the tube. The argon is seeded with electrons by momentarily connecting a Tesla coil to the tube and the plasma forms inside the tube. The ions in the gas tend to flow in a circular path around the lines of force of the

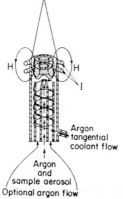

Fig. 2.2. Plasma torch arrangement. (From Ref. 9.)

oscillatory magnetic field and the resistance to their flow produces the heat. It is a mechanism analogous to the production of heat in a metal cylinder placed in an induction furnace. To avoid melting the silica tube a flow of argon is introduced tangentially in the tube and this centres the plasma away from the walls of the tube. The plasma is formed in the shape of a toroid or doughnut and the sample is introduced as an aerosol through the middle of the toroid. The arrangement of the plasma 'torch' is shown in Fig. 2.2. The hottest part of the plasma is in the ring around the centre of the toroid where temperatures of about 10 000 K are achieved. Through the centre of the toroid where the sample is introduced the temperature is somewhat lower and the sample is subjected to temperatures of about 7000 K. From the very hot region in the plasma and just above it a continuum is radiated because of the high electron density. Above this region the continuum emission is reduced as the temperature falls and the spectral lines of the elements in the sample may be observed.

There are now at least six companies producing instruments for multi-element analysis using ICAP. A picture of the one made by Jarrell-Ash is shown in Fig.

Fig. 2.3. Multi-element analyser using ICAP. (Courtesy Jarrell-Ash Division, Fisher Scientific Co.)

2.3. In most of these commercial instruments the spectrum observed is resolved in a spectrometer and the relative intensities of selected lines of the elements are fed to a small computer. This performs the calculations of concentrations and, typically, up to 50 elements can be determined in a sample. The time required for an analysis is one minute or less and hence, with automated sample handling, a large number of samples can be processed. The characteristic property of the emission analysis that makes feasible the automated analysis of a variety of samples is that the dynamic range of concentration that can be measured is large. A single calibration curve can cover changes in concentration of five orders of magnitude. Typical detection limits obtained with ICAP analysis are shown in

Table 2.5. When the technique was first introduced there were claims that it was free from interferences. However, it is now recognized that interferences do exist and must be corrected for. One advantage of the computer calculation of concentration is that if it is recognized that one of the elements being determined interferes with another, a correction can be made via the calculation programme.

In most of the work using the ICAP technique which has been reported to date the sample has been processed in solution, most commonly by a pneumatic nebulization process similar to that used in atomic absorption technology. This process is an inefficient way of introducing the sample, and other techniques which result in greater amounts of the sample entering the plasma have been developed. Ultrasonic nebulization, spraying into a heated chamber, chemical generation of vapours and evaporation from a filament are some of the varieties of methods used. References to these techniques may be found in the review by Greenfield et al.[8] Because of the high temperature of the plasma solid samples introduced as a fine powder will also vaporize and this is one possible route for elemental analysis of solid samples. An alternative way of handling solid samples is to vaporize them in a graphite tube or on a tantalum wire, such as is used in flameless atomic absorption, and to sweep the vapour into the plasma. This also offers a way of increasing the amount of a liquid sample that is fed to the plasma.

Atomic fluorescence

The process by which atomic fluorescence is used for analysis is shown schematically in Fig. 2.4. A sample is converted to a cloud of atoms in a gaseous state in

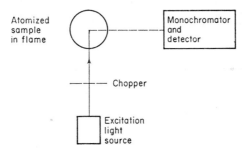

Fig. 2.4. Atomic fluorescence schematic.

similar ways to those used in atomic absorption spectroscopy. This cloud is illuminated by radiation that the atoms will absorb and thus be raised to an excited state. The atoms then fall from the excited state to a less excited state and emit radiation in so doing. In order to limit interference from the incident radiation this emitted radiation is measured at right angles to the incident radiation via a monochromator or other wavelength-selective device. The amount of radiation emitted is a function of the number of atoms in the radiation absorption zone, the intensity of the incident radiation, the efficiencies of the various processes, the geometry of the system, etc.

Although there are 14 possible processes by which an atom can absorb radiation energy and re-emit it as fluorescence radiation, only a few of these are of analytical importance. Five of the possible processes are shown diagrammatically in Fig. 2.5. The process most used in analytical work is that of 'resonance fluorescence' (a). The atom absorbs energy to become excited and then emits radiation of the same wavelength as it falls back to its original ground state. In the 'direct-line' process the atom absorbs radiation to go to an excited state and then emits radiation in going to a state that is not the original level. In the simple direct-line process shown in (b) the atom goes to the 2nd excited state and then emits radiation in falling back to the 1st excited state. An example of this is the emission of thallium at 535.0 nm after absorbing energy at the 377.6 nm thallium line. In the thermally assisted direct-line process (c) the atom

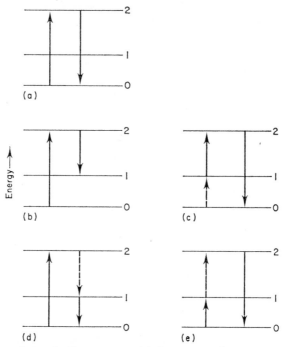

Fig. 2.5. Types of atomic fluorescence. (a) Resonance fluorescence. (b) Direct-line fluorescence. (c) Thermally assisted direct-line fluorescence. (d) Stepwise-line fluorescence. (e) Thermally assisted stepwise-line fluorescence. Solid lines—radiational processes; dotted lines—non-radiational processes.

absorbs energy by a non-radiational process, such as collision, to be in the 1st excited state. It then absorbs radiation energy to go to the 2nd excited state and re-emits energy in falling from the 2nd excited state to the ground state. The stepwise-line processes, shown in (d) and (e), are less used for analytical work than the other three. In these processes the atom absorbs radiation to go to an excited state, and either loses energy or absorbs energy by a non-radiation process to go to another excited state. It then emits radiation in going back to the original ground state. An example of stepwise line fluorescence is when sodium is excited with the 330.3 nm line and emits radiation at the 584.0 nm line.

The source of the excitation radiation is of great importance in atomic fluorescence since the intensity of the radiation emitted, and measured, is proportional to the intensity of the source. Hence it is possible to achieve greater sensitivity by increasing the intensity of the source. The source may be a line source such as a microwave-excited electrodeless discharge lamp or a continuum source such as a xenon arc lamp. Hollow cathode lamps have been used as sources but the regular lamp as used in atomic absorption does not give enough light and it is necessary to use a special high-radiance lamp or to pulse the lamp with an intermittent high current. The use of laser sources appears to be the most attractive route to high sensitivity fluorescence measurements.

The atomization source may be either a flame or a flameless device. The principles involved are the same as in atomic absorption but the geometry is different because of the 'right-angle' nature of the excitation and emission radiation. A disadvantage of the flame is that CO, CO_2 and N_2 act as radiation quenchers to reduce the atomic fluorescence. The effect of CO and CO_2 can be avoided by using non-hydrocarbon fuels such as hydrogen, and the effect of nitrogen can be overcome by sheathing the flame in argon. However, for practical use, air–acetylene or nitrous oxide–acetylene flames must be used to overcome effects of chemical interference in the atomization process. The background emission is reduced by separating the secondary reaction zone of these flames with a concentric stream of argon.

In many atomic fluorescence applications a non-dispersive optical system may be used because of the specificity of the fluorescence radiation. This makes the cost much lower than the corresponding monochromator system and is particularly useful in multi-element application. It should be noted that in many of the systems proposed for multi-element analysis the different excitation sources are pulsed to give intermittent radiation; the detectors of the emitted radiation are synchronized to their specific elements' source. Hence the determination is not a simultaneous one which would require resolving the spectrum of all the elements, but a repetitive sequential one over a very short time span.

The sensitivities of the various atomic spectroscopy techniques have been compared in a number of articles. While some elements are more sensitive by one technique than another, over-all the sensitivities are about the same. A comparison of detection limits recently reported for atomic absorption, atomic fluorescence and atomic emission (by ICAP and by flame) is shown in Table 2.5. Since there are no commercially available instruments specifically designed for atomic fluorescence there are no instruction manuals similar to the 'cook-books' available for atomic absorption. There have been a number of reviews published, one comprehensive one by Browner,[10] and the recent book by Kirkbright and Sargent[119] covers applications of atomic fluorescence.

MOLECULAR ABSORPTION SPECTROPHOTOMETRY

Most of the metals that are of environmental concern can be determined by reacting the metal ion in solution with an organic complexing agent. The concentration of the resulting compound is determined by a photometric measurement at the appropriate wavelength. Since most of the complexes used absorb in the visible region, in many cases it is feasible to carry out the analysis with a visual comparison of the colours formed in the samples and standards. The attraction of these methods, particularly for water samples, is that they can be carried out with a minimum of equipment. Another, perhaps alternative, attraction is that the analysis can be done in readily available automated equipment such as the continuous-flow Technicon Auto Analyzer® or in a discrete sample analyser.

TABLE 2.6
Reagents used for spectrophotometric determination of metals

Metal	Reagent	Wavelength/nm	Minimum detectable concentration/μg l^{-1}
		Automated methods	
Fe[28]	TPTZ (2,4,6 Tripyridyl-*s*-triazine)	600	20
Cr[29]	Diphenylcarbazide	550	4
B[30]	Carminic acid	600	20
Cu[31]	Bis-cyclohexanoxaldihydrazone	590	4
Mn[31]	Formaldoxime	480	10
		Reagents from *Standard Methods*[32]	
Ag	Dithizone	620	2
Al	Eriochrome cyanine R	535	6
As	Silver diethyldithiocarbamate	535	5
B	Curcumin	540	200
Be	Aluminon	515	5
Cd	Dithizone	515	10
Cr	Diphenylcarbazide	540	5
Cu	Neocuproine	457	6
Fe	Phenanthroline	508	20
Hg	Dithizone	490	2
Mn	Persulfate-periodate	525	50
Ni	Heptoxime	445	20
	Dimethylglyoxime	445	20
Pb	Dithizone	520	5
Se	Diaminobenzidine	420	1
V	Dithizone	620	2
Zn	Dithizone	535	1
	Zincon	620	20

The aim in the selection of the colour-forming agent is to use one which is selective to the metal being determined, produces a complex whose absorption peak is well separated from that of the reagent, and which has a high extinction coefficient, i.e. produces an intense colour. Unfortunately this is possible in only a few cases, the metals which can be determined in natural water samples merely by the addition of reagents to their solution being Fe, Cu, Cr, B and Mn. These are the metals for which automated methods are available. In Table 2.6 are shown the reagents used in the automated methods. Also shown in Table 2.6 are the reagents recommended in *Standard Methods*[32] for the colorimetric determination of a number of metals, together with the wavelengths used for the measurements and the minimum detectable concentrations.

Many of the complexing agents will react with many metals. Dithizone, for example, will complex 20 different metals, and the analytical procedures include a separation step to isolate the desired metal before the colour formation is carried out. The separation process in which there appears to be the most interest now is ion-exchange chromatography. In this the solution containing the mixture of metals is passed through an ion-exchange column under conditions where many of the interfering ions will not be adsorbed but the desired ion plus some others will be. By choice of elution conditions the metals retained on the resin can be desorbed at different times and the desired metal can be isolated in a certain fraction of the eluant. Thus Matsui describes methods for the determination of zinc[11] and manganese[12] where both use zincon (2-carboxy-2'-hydroxy-5'-sulfoformazylbenzene) as the complexing reagent but they are separated from the other metals by ion-exchange chromatography. Similarly cobalt is determined with nitroso-R salt[13] and chromium with PAR (4-(2-pyridylazo)-resorcinol).[14] These are both reagents which react with many metals.

A particular application of colorimetric analysis is used in the ring-oven. This is a technique developed to make separations and semi-quantitative determinations on micro quantities of material. It has been used in determining the composition of air particulates. The sample solution is spotted on an absorbent paper and spreads out in a circle until it reaches a heated annulus (or ring). There the solvent evaporates and the solute is deposited. Addition of solvent to the centre of the circle transfers all the solute to the heated ring. Micro separations are made by precipitating some of the components so that they will no longer migrate with the solvent. The materials in the ring are identified by colorimetric spot test reagents, and their concentrations are estimated by comparison with standard ring colours. This technique is described by Weisz.[15]

ELECTROCHEMICAL METHODS

Electrochemical methods of analysis have the advantage that the species of the particular metal being analysed can be characterized. The methods will distinguish, for example, between the metals present in an ionic state and those that are complexed. Hence the methods are of particular value in analysing water samples where the species of the metals is of concern. The techniques described are polarography and anodic stripping voltammetry.

Polarography

In classical polarography a dropping mercury electrode (DME) is used. This consists of a capillary connected to a mercury reservoir. The mercury flows through the capillary and out of the end in a series of small drops. The sample is taken, a supporting electrolyte is added as necessary, and the DME and a reference electrode are placed in the solution. There is a means of applying a variable potential between the electrodes and of measuring the current flow. The potential between the electrodes is slowly increased. When a potential is reached where the metal of interest is, for example, reduced at the electrode surface, the current starts to increase. On further increase in voltage the current rises to a point where the amount of the metal being reduced and hence the current, is

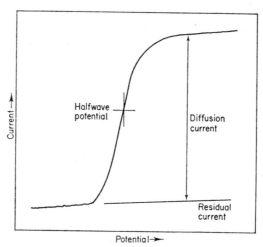

Fig. 2.6. Typical polarograph curve.

limited by the rate at which the reducible species can diffuse through the liquid to the surface of the electrode. This current, the diffusion current, is a measure of the concentration of the metal. The potential at which the reduction occurs is characteristic of the metal in the particular electrolyte system being used. Metals which are complexed will not reduce at the same potential as the free metal ion. For convenience the potential at which the reduction occurs is measured at the halfway point in the current increase and is known as the halfwave potential (see Fig. 2.6).

The current–voltage curve is obtained over several minutes, the forming and dropping off of the mercury drops in the DME gives the familiar saw-tooth outline to the curve. A typical curve is shown in Fig. 2.6. The sensitivity of the method is dependent on the accuracy with which the faradaic current (i.e. that producing the reduction) can be measured. It is possible to measure very low currents, but in the system described the mercury drop and the solution around it forms an electrolytic condenser which charges as the drop forms. This capacitance current cannot be distinguished from the faradaic current. In addition, the residual current and the current due to impurities in the supporting electrolyte are often difficult to enumerate. These factors limit the use of the simple polarograph to analysis of the solutions of 10^{-5} M or greater (~ 0.5 mg l^{-1} for Cu and Zn).

Improvements are made in sensitivity by using differential polarography. In this technique two cells as nearly identical as possible are coupled together with a common applied voltage source. The difference in current flow between the two cells is measured, and this gives opportunity for many of the sources of non-signal current to be cancelled out. In derivative polarography the rate of change of current with the applied voltage is plotted instead of the change in current. This gives better resolution of the waves from different elements since the rate of change of current is a maximum at the halfwave potential.

In pulse polarography a pulse of about 40 mV is applied to the electrode. Immediately there is a sharp increase in current to charge the double layer. This current dies away, typically in about 40 ms, and the faradaic current corresponding to the 40-mV increase in potential can then be measured. Alternatively with an oscilloscope to measure rapid changes in current the full voltage sweep can be applied in a fraction of the lifetime of a single mercury drop. Other techniques apply an alternating potential or current to the mercury drop, either as a sine wave or square wave. In the 'rapid potential application' methods, diffusion does not limit the process any more; this gives an increase in sensitivity. With these sophisticated techniques it is possible to use polarography to measure levels down to 10^{-8} M (~ 0.5 μg l^{-1} for Cu and Zn).

Anodic stripping voltammetry

A method that makes possible a very sensitive determination of some materials in water is stripping voltammetry. In the form described here, for the determination of metals, it is anodic stripping, although cathodic stripping is also used.

There are two processes carried out. The metal ions in solution are first plated out on an electrode, either completely or under controlled conditions, so that a known fraction is isolated. The metal on the electrode is then determined by applying an increasing reverse potential to the electrode and recording the current in a way similar to that used in the polarograph. There is a current peak as the metal is oxidized from the electrode into the solution. The potential at which the current peak is observed is specific to the metal; the height of the current peak is proportional to the concentration.

In its simpler form the metals are plated on to a hanging mercury drop electrode and stripped by applying a uniformly increasing potential. This is called linear ramp stripping. The use of a hanging mercury drop imposes some limitations on the process. The surface area of the drop is small so that the deposition rate is low; the volume is small so that the metal on stripping has to diffuse from inside the drop, giving broad peaks; and the stirring rate in the solution is limited by the mechanical stability of a hanging mercury drop. The use of a mercury film on a solid carbon electrode overcomes most of these limitations. In the stripping step there is the same problem as was described above for polarography, namely the distinguishing of the faradaic current from the capacitance current. It is not as severe a problem, however, as the concentrations at the electrode are much higher than in polarography. For best sensitivity with linear ramp stripping the voltage gradient should be high, but this also gives rise to large capacitance currents. The technique of applying voltage pulses to the stripping ramp is used as described above. In the stripping operation, after each pulse the electrode returns to a lower voltage and may be cathodic to the metal ions which have just come off the electrode and have not diffused away into the solution. These ions may then replate on the electrode and make another contribution to the faradaic current on the next pulse. This improves the sensitivity of the pulse mode of operation. Also used is the addition to the stripping voltage of an a.c. component. Phase measurements will separate the faradaic and capacitive currents since these occur at different phases relative to the applied a.c. voltage.

Using a mercury-coated graphite electrode, bismuth and antimony have been measured in sea water in the ranges 0.02–0.9 μg kg^{-1} and 0.2–0.5 μg kg^{-1} respectively.[16] Cadmium, copper, lead and zinc are commonly determined in water samples down to 0.1 μg l^{-1} levels.

Anodic stripping is limited to those elements which are readily reduced to the metal and re-oxidized. These are Ag, As, Au, Ba, Bi, Cd, Cu, Ga, Ge, Hg, In, K, Mn, Ni, Pb, Pt, Rh, Sb, Tl and Zn. For the metals which have an oxidation potential anodic to mercury, e.g. Au, Hg and Ag, a solid electrode rather than a mercury electrode must be used.

NEUTRON ACTIVATION ANALYSIS

Neutron activation analysis provides an accurate and sensitive means of determining a large number of elements in a wide variety of samples. It has the advantage that the chemical form of the element being analysed and the physical form of the sample do not have much effect and hence the majority of environmental samples may be processed without any chemical pretreatment. It is also a multi-element analytical technique. In the past it has had the disadvantage that the facilities for carrying out the neutron irradiation were very expensive and limited to a few university departments and government laboratories. However, over the last few years the cost of small nuclear reactors has decreased so that they are now more widespread among universities. In addition, other neutron sources are becoming available. One of these sources is the use of accelerators; the system most commonly used to produce neutrons is one in which deuterons are fired at a target consisting of tritium diffused in a metal (usually Pd). This produces neutrons by the $^3H(d, n)^4He$ reaction and these neutrons have an energy of about 14 MeV. Another source is californium-252. This material is being made available by the U.S. Atomic Energy Commission. It is a self-fissioning material and its ready portability makes possible the development of facilities for mobile activation analysis. The lowering of cost and the availability of neutron sources other than reactors has resulted in increased use of neutron activation as an analytical tool. At the present time a significant amount of data that is appearing in the literature on the occurrence of elements in the environment has been obtained by neutron activation analysis.

For the majority of analyses the neutrons used are thermal neutrons, i.e. neutrons which have energies of about 0.02 eV, so that the samples are normally irradiated in the 'moderated' part of a nuclear reactor. The analysis operates by the fact that an element subjected to bombardment by thermal neutrons will absorb some of these neutrons to form an isotope of the same element by an (n, γ) reaction, i.e. a neutron is absorbed and γ radiation is emitted.

$$^A X + n \rightarrow {}^{A+1}X + \gamma \text{ radiation}$$

where A is the mass number of the element and $A + 1$ is the mass number of the isotope formed. The isotope formed is generally radioactive and will decay by β emission (i.e. loss of an electron) to an element of atomic number higher by one. The element formed by this decay is generally in an excited state and, immediately after the emission of the β particle, energy is lost in the form of γ radiation. The energy distribution of this γ radiation is characteristic of the particular decay process.

In practice the sample is sealed in a container and irradiated for a certain time, which may range from a few minutes to many hours. Before irradiation a solid sample may be merely ground and homogenized; liquid samples and small samples may be irradiated as received. Since the amount of the isotope produced is dependent both on the amount of the original element present and on the neutron flux, a standard containing the element(s) to be determined should be irradiated at the same time, to determine the neutron dosage that the sample receives. After irradiation the amount of the isotope and hence the amount of the element originally present may be determined from the γ-ray spectrum of the sample. Alternatively a chemical separation may be carried out to isolate the isotope and to determine its concentration by measuring its radioactivity.

In the determination of the γ-ray spectrum the irradiated sample is placed close to a crystal. The photons displace electrons in the crystal elements and the light photons, given off as these displaced electrons return to their normal state, enable a detector to measure the distribution of energy in the incident radiation. The crystals used until recently (since 1948 when they were first introduced) were made of sodium iodide and were not able to resolve the γ energy very well. Currently, lithium-drifted germanium, Ge(Li), crystals are being used which have much better resolution than the NaI(Tl) crystals. These make viable the analysis of mixtures of isotopes with complex spectra, such as those produced in environmental samples. (For further discussion see Chapter 5.)

The alternative to a γ-ray spectrum determination of the elements is to carry out a chemical separation. A known 'macro' amount of the element to be determined is added to the irradiated sample and some chemical process, such as an oxidation or reduction, is carried out to ensure that the radioactive isotope is completely mixed with the added stable isotope. A separation is then carried out to isolate the particular element, e.g. in the case of mercury it may be separated as the sulfide or as elemental mercury. The activity of the separated material is then counted. By a chemical determination of the recovery of the stable 'carrier' added at the beginning of the separation the efficiency of the separation can be calculated and the necessary factor applied. In a lot of the early work reported, chemical separations were carried out because of the difficulty and cost of determining and resolving γ-ray spectra. Normally separations are now only carried out if there is a need to obtain a much increased sensitivity or if there is interference from a highly active nuclide.

Perhaps the most significant source of error in the technique is that if the reaction assumed is an (n, γ), other reactions may occur with other elements to produce the same isotope. For example, in the case of phosphorus determination

TABLE 2.7
Detection limits by neutron activation (from Yule, Ref. 17)

Element	Isotope used	Halflife	Detection limit/μg
Ag	Ag^{108}	2.3 min	0.0034
Al	Al^{28}	2.3 min	0.0049
As	As^{76}	1.1 d	0.0015
B	—	—	—
Ca	Ca^{49}	8.8 min	1.5
Cd	Cd^{111m}	49 min	0.0039
Co	Co^{60m}	10.5 min	0.0032
Cr	Cr^{51}	27.8 d	0.45
Cu	Cu^{64}	12.8 h	0.001
Fe	Fe^{59}	45.1 d	100
Hg	Hg^{197m}	24 h	0.083
Mg	Mg^{27}	9.5 min	0.14
Mn	Mn^{56}	2.58 h	0.000036
Mo	Mo^{99}	66 h	0.047
Na	Na^{24}	15 h	0.0029
Ni	Ni^{65}	2.56 h	0.180
Pb	Pb^{207m}	1 s	~200
Sb	Sb^{122}	2.8 d	0.0020
Se	Se^{81m}	61 min	0.021
Sn	Sn^{123}	41 min	0.013
V	V^{52}	3.76 min	0.00038
Zn	Zn^{69m}	13.8 h	0.083

in sea water the reaction with the chloride of $^{35}Cl(n, \alpha)^{32}P$ gives the same isotope as the $^{31}P(n, \gamma)^{32}P$ reaction. Another interference is where fast neutrons may react with another element to produce (n, p) reactions such as $^{60}Ni(n, p)^{60}Co$

TABLE 2.8
Detection limits and nuclide properties in analysis of air particulates (from Zoller and Gordon, Ref. 18)

Element	Target isotope	Isotopic abundance/%	Product nuclide	Halflife	Best time after irradiation for counting	Detection limit/μg m^{-3}
Short runs						
Al	^{27}Al	100	^{28}Al	2.3 min	0–20 min	0.001
Br	^{79}Br	50.56	^{80m}Br	4.4 h		0.002
			^{80}Br	17.6 min	1–4 h	—
	^{81}Br	49.44	^{82}Br	35.3 h	4–6 d	—
Ca	^{48}Ca	0.185	^{49}Ca	8.8 min	0–20 min	1.0
Cl	^{37}Cl	24.47	^{38}Cl	37.3 min	20–60 min	0.2
Cu	^{65}Cu	30.9	^{66}Cu	5.1 min	0–20 min	0.02
Mn	^{55}Mn	100	^{56}Mn	2.58 h	1–4 h	0.001
Na	^{23}Na	100	^{24}Na	15.0 h	1 d	0.001
V	^{51}V	99.75	^{52}V	3.77 min	0–20 min	5×10^{-4}
Long runs						
Ba	^{130}Ba	0.1	^{131}Ba	12 d	2 weeks	10^{-4}
Ce	^{140}Ce	88.5	^{141}Ce	33 d	2–4 weeks	10^{-5}
Co	^{59}Co	100	^{60}Co	5.26 y	$\geq$1 month	10^{-6}
Cr	^{50}Cr	4.31	^{51}Cr	27.8 d	1 month	—
Cs	^{133}Cs	100	^{134}Cs	2.05 y	$\geq$1 month	—
Eu	^{151}Eu	47.8	^{152}Eu	12.5 y	$\geq$1 month	10^{-5}
Fe	^{58}Fe	0.31	^{59}Fe	45 d	1 month	0.001
Hf	^{180}Hf	35.2	^{181}Hf	42.5 d	1 month	10^{-5}
La	^{139}La	99.9	^{140}La	40.2 h	4–6 d	5×10^{-5}
Lu	^{176}Lu	2.6	^{177}Lu	6.7 d	2 weeks	10^{-5}
Ni	^{58}Ni	67.8	$^{58}Co(n,p)$	71 d	$\geq$1 month	0.01
Sb	^{121}Sb	57.25	^{122}Sb	2.8 d	1 week	10^{-5}
	^{123}Sb	42.75	^{124}Sb	60 d	$\geq$1 month	
Sc	^{45}Sc	100	^{46}Sc	84 d	$\geq$1 month	2×10^{-5}
Se	^{74}Se	0.87	^{75}Se	120 d	$\geq$1 month	2×10^{-5}
Sm	^{152}Sm	26.6	^{153}Sm	47 h	4–6 d	5×10^{-5}
Th	^{232}Th	100	^{233}Th	22 min		10^{-5}
			^{233}Pa	27 d	2–4 weeks	
Yb	^{168}Yb	0.14	^{169}Yb	32 d	1 month	10^{-5}
Zn	^{64}Zn	48.9	^{65}Zn	245 d	$\geq$1 month	2×10^{-5}

which yield the same product as the thermal neutron reaction $^{59}Co(n, \gamma)^{60}Co$. However, since in a reactor the fast neutron flux is generally much lower than that of the thermal neutrons and the fast-neutron cross-sections are also much lower than the thermal-neutron cross-sections, this type of interference is not a problem unless the concentrations of the various elements are different by several orders of magnitude.

Yule[17] has published a list of the detection limits obtained in thermal neutron activation. Some of these are shown in Table 2.7. It should be noted, however, that this data was obtained by irradiating single elements for a standard time and calculating the smallest amount of the element that would give a count that could be distinguished from the background. Hence inter-element interferences were not considered. Zoller and Gordon[18] have used γ-ray spectrometry to analyse atmospheric aerosols and they have determined the detection limits for 24 of the 26 elements analysed. These detection limits and the properties of the nuclides used are shown in Table 2.8. In this work distinction was made between the nuclides with long and short halflives by carrying out two separate irradiations. The filter containing the particulates was halved and one half irradiated for about five minutes to activate the short-lived nuclides. The γ-spectrum of this half was then examined at various times over the following days as the very short-lived nuclides decayed. The other half of the filter was irradiated for several hours and then allowed to 'cool' for several days so that the long-lived activity could be measured. Pillay and Thomas[19] have analysed air particulates taken in the Buffalo, N.Y., area for over 20 elements. The irradiation, decay periods, etc., used by them are shown in Table 2.9.

TABLE 2.9

An irradiation–decay scheme used for the analysis of air pollutants (from Pillay and Thomas, Ref. 19)

Irradiation period	Thermal neutron flux/n cm^{-2} sec^{-1}	Decay period	Elements analysed
5 min	5×10^{12}	0–10 min	Al, V, I, Cl, Mn, Na
2 h	5×10^{12}	8–12 h	Na, K, Cu, Au, Ag, Br, Hg and rare earths
24 h	3×10^{13}	5–7 d	Fe, Zn, Cr, Ag, Co, Ni, Mo and rare earths
20 h	3×10^{13}	5–10 d	Se (after chemical separation)
0.01–0.03 s	10^{14}–10^{15}	0.1–0.4 s	Pb

The element that is difficult to analyse by neutron activation is lead. The only long-lived (n, γ) product is 3.3 h ^{209}Pb. This has a very low formation cross-section and decays by β emission only. Pillay and Thomas[19] used 0.8 s ^{207m}Pb

with very short irradiation and fast transfer to the spectrometer. An alternative for lead is to activate by particle bombardment.

One advantage that neutron activation possesses is that samples can be irradiated without any chemical pretreatment, thus minimizing the risk of contamination. This is particularly attractive for solid samples. In the case of water samples, however, where it is desired to measure the trace elements, there are often advantages in carrying out a separation–concentration step before the irradiation. In this way it is possible to separate the elements such as sodium and chlorine which may be present in the sample in relatively large amounts and which give highly active nuclides on irradiation. The separation processes used are similar to those carried out for atomic spectroscopy. In one example, Fujinaga et al.[20] analysed V, Al, Cu, Mo, Zn and U in natural water samples using organic co-precipitation as a preconcentration method. They report the levels of these elements in several natural waters at the $\mu g\, l^{-1}$ level.

This discussion has been directed towards thermal neutron reactions. The same techniques are applied when fast neutrons are used; the reactions proceeding are generally ones of particle emission rather than γ-ray emission. In Table 2.10 is shown some of the fast neutron reactions that are of interest in analysing environmental samples.

TABLE 2.10
Some fast neutron reactions

Element	Target isotope	Reaction	Cross-section for 14 MeV Neutrons/barns	Product nuclide	Halflife
Ag	^{109}Ag	$n,2n$	0.700	^{108}Ag	2.3 min
Cd	^{111}Cd	$n,n'\gamma$	—	^{111m}Cd	49 min
Co	^{59}Co	n,α	0.039	^{56}Mn	2.6 h
Cu	^{63}Cu	$n,2n$	0.550	^{62}Cu	9.8 min
Cr	^{52}Cr	n,p	0.078	^{52}V	3.8 min
Fe	^{54}Fe	n,p	0.373	^{54}Mn	310 d
	^{56}Fe	n,p	0.103	^{56}Mn	2.6 h
Ni	^{58}Ni	n,p	0.237	^{58}Co	71 d
Pb	^{204}Pb	$n,n'\gamma$	0.160	^{204m}Pb	67 min
Se	^{77}Se	$n,n'\gamma$	—	^{77m}Se	18 s
V	^{51}V	n,p	0.027	^{51}Ti	5.8 min

X-RAY FLUORESCENCE

In X-ray fluorescence measurements the sample is subjected to an excitation that displaces an electron from one of the inner shells of the atom. On the resulting rearrangement of the electrons, to return to a stable configuration, energy is released as radiation. The energy changes that occur in this process are in the X-ray part of the electromagnetic spectrum. Since the electron displacement is in the inner shells, the chemical state of the element is not important. The physical state is important only in that the samples and the standards used in calibration can be exposed to the excitation and measurement of the X-ray spectrum in a reproducible manner.

In the case of air particulates the filters used to collect them are used directly, as are filters used to separate particulate matter in water samples. Solid samples such as vegetation are dried and pressed into pellets.

The excitation may be made by X-rays or γ-rays from radionuclides, by X-rays from tubes or by charged particles such as electrons, protons, or α-particles.

The determination of the X-ray spectrum may be made in two ways: by wavelength dispersion or by energy dispersion. Wavelength dispersion is the older technique. The X-rays induced in the sample are collimated, diffracted by a crystal and detected by a counter placed at the required geometry to the crystal. By turning the crystal the wavelength distribution is obtained. The crystal material must be selected such that it will diffract the characteristic X-rays for the element being analysed. The technique is sensitive, but for multi-element analysis it is time-consuming because of the multiple scanning needed.

The newer technique, which has the advantage that it will more easily handle some multi-element analyses, is that of X-ray energy spectrometry. The induced X-rays are directed at an Si(Li) detector connected to a multi-channel analyser,

TABLE 2.11

Detection limits for X-ray fluorescence analysis of air particulates (from Gilfrich et al., Ref. 24)

| Element | Wavelength Dispersion | | Energy Dispersion | |
	ng cm^{-2} Filter	μg m^{-3} Air[a]	ng cm^{-2} Filter	μg m^{-3} Air[a]
Al	50	0.012	—	—
Br	130	0.032	—	—
Ca	17	0.004	53	0.013
Cu	29	0.007	—	—
Fe	18	0.004	46	0.011
K	11	0.003	4	0.001
Pb	160	0.040	42	0.010
S	8	0.002	220	0.055
Se	90	0.022	31	0.008
V	20	0.005	34	0.008
Zn	31	0.008	42	0.010

[a] Data from filter area transposed assuming 4 m^3 air per cm^2 filter.
Excitation by X-ray tube. Wavelength Dispersion—Rh Tube; Energy Dispersion—W Tube–Ni Foil.

a system similar to that used in determining the γ-ray spectrum in neutron activation analysis. The resulting energy spectrum is processed to enable a determination of the elements present to be made. Energy-dispersive analysis of X-rays has been discussed by Russ[21] and also briefly by Porter and Woldseth.[22]

An X-ray fluorescence analyser is available in a simple form such as the Applied Research Laboratories Limited N900 instrument described by Gamage.[23] This consists of isotope sources and a small X-ray tube to irradiate the samples, and a detector system that enables the energy levels of the resulting X-rays to be scanned manually. This allows the analysis of material for a single element at a time by choice of isotope, detector and energy channels.

A comparison of means of excitation and detection in the analysis of air particulates has been made by Gilfrich et al.[24] One of the conclusions reached in this study was that the current energy dispersive techniques did not have enough resolution to separate the Kα line of one element from the Kβ line of the element of one lower atomic number in the range of sulfur to nickel. Hence for determination of these elements in air pollution samples, multi-channel wavelength dispersion spectrometers are required.

A comparison of trace element determination in simulated and real air particulates by various X-ray fluorescence methods and by other non-X-ray methods was carried out by Camp et al.[25] The elements measured in this work were Al, S, K, Ca, Ti, Mn, Fe, Cu, Zn, Se, Br and Pb. The results showed that energy dispersive X-ray spectrometry is a viable way to measure these elements in air particulates. Gilfrich et al.[24] determined the detection limits of several elements in their study. They extrapolated the values they found to what would be reasonable using commercially available high-powered equipment. These extrapolated values are quoted in $ng \ cm^{-2}$ of filter. For easier comparison with other data these have been converted to $\mu g \ m^{-3}$ of air and both values are shown in Table 2.11.

One approach to the analysis of trace metals in water has been to pass the water through a filter containing an ion-exchange or chelating resin. This extracts the metal ions and the filter is then analysed by X-ray fluorescence. This method is discussed by Carlton and Russ.[26] Another similar approach is to precipitate the metals with an organic carrier, filter the precipitate and similarly examine it.[27] Levels of heavy metals at fractions of $\mu g \ l^{-1}$ are measured with this technique.

CURRENT DEVELOPMENTS

In Table 2.12 are given references to recently published methods for the determination of metals in environmental samples by the techniques described above.

TABLE 2.12
References to recently published analytical methods

Technique	Ag	Al	As	B	Be	Ca	Cd	Co	Cr	Cu	Fe	Hg
Water												
Preservative[a]	(e)	(a)	(c)	—	(a)	(a)	(a)	(a)	(a)	(a)	(a)	(b)
Atomic absorption	33		34, 35, 36									37
Atomic fluorescence			43								48	44
Molecular absorption spectrophotometry			45, 46	47								
Electrochemistry							51, 52			51, 52	53, 55	
Neutron activation	53	53	53, 54, 55			53, 55	53, 55	53, 55	53, 55	53, 55	53, 55	53, 55
X-ray fluorescence							53, 56			26, 57, 58	26, 57, 58	57, 58
Sediment and Soil												
Atomic absorption		59	35, 60, 61		59, 62	59, 63	59, 64, 65	59	59, 66	59, 66	59, 66	67
Atomic emission		70		71		70					70	
Atomic fluorescence			43		73						74	72
Molecular absorption spectrophotometry				47							77	
Neutron activation	77	77	77, 78, 79			77	77	77	77	77	80, 81, 82	77
X-ray fluorescence	80		80, 81				80		80	80, 82		
Air												
Atomic absorption		83	40		84	83	83	83	83	83		
Atomic emission			86			86				86		
Neutron activation	77, 87	77, 88	77, 88			77, 88	77	77, 88, 89, 90	77, 78, 90	77, 88	77, 90	77
X-ray fluorescence						91		91	91	91	91, 92	
Biological Samples												
Atomic absorption	93	94	35, 36, 60, 95, 96		62		97, 98	97, 98		97	97	100
Atomic emission	107				107		107	107	107	107	101	
Atomic fluorescence											74	72
Molecular absorption spectrophotometry			43									
Electrochemistry			108									
Neutron activation	111, 112	75	75, 111, 112, 113			112	75, 111, 112, 113, 114	75, 111, 112, 113, 114	75, 111, 112, 113, 114	75, 111, 112, 113, 115	75, 111, 112, 113	75, 111, 112, 113, 115
X-ray fluorescence	116	116				116				117	117	

TABLE 2.12 contd.

Technique	Mg	Mn	Mo	Na	Ni	Pb	Sb	Se	Sn	V	Zn
Water											
Preservative[a]	(a)	(a)	(a) 38	(a)	(a)	(a) 39	(c) 40, 43	(d) 40, 42, 51, 43	(a)	(a)	(a)
Atomic absorption											
Atomic fluorescence										49, 50	
Molecular absorption spectrophotometry											
Electrochemistry	53	53	53, 55	53, 55	55	51, 52	53, 54, 55	53, 55		53	51, 52, 53, 54
Neutron activation					26	26, 56, 57					26, 56, 57, 58
X-ray fluorescence											
Sediment and Soil											
Atomic absorption	59, 63	59, 66	59, 68, 38	59	59, 66	59, 66			69	59	59, 65, 66
Atomic emission	70	70		70							
Atomic fluorescence							43	43			
Molecular absorption spectrophotometry											
Neutron activation	77	77	77, 78	77	77		77, 78	77, 79		75, 77	77
X-ray fluorescence		80, 82			80	80, 82	80	80	80	80	80, 81, 82
Air											
Atomic absorption	83	83			83	83, 39			85		83
Atomic emission					86						86
Neutron activation	77, 88, 90	77, 88	77	77, 88	77, 88	88	77, 89	77		77, 88	77, 88
X-ray fluorescence		91			91	92				91	92
Biological Samples											
Atomic absorption	101	97, 102			97	39		103, 104, 105, 106			97
Atomic emission	107			107	107				107	107	107
Atomic fluorescence								43	43	43	
Molecular absorption spectrophotometry								109, 110			
Electrochemistry											
Neutron activation	112, 113		75, 111, 113	111, 113; 111, 112, 113	113	111, 112	75, 111, 112, 115	113	111	114	75, 111, 112, 113, 114
X-ray fluorescence	116		116					118			117

[a] (a) 0.2% conc. HNO_3; (b) 1% H_2SO_4 + 0.05% $K_2Cr_2O_7$; (c) 0.2% H_2SO_4; (d) 0.2% H_3PO_4; (e) 0.4% Na_2EDTA.

This is not intended as an exhaustive review, but is limited to methods published in the last two or three years in widely distributed journals in English.

REFERENCES

1. S. Slavin, W. B. Barnett and H. L. Kahn, *At. Absorpt. Newsl.* **11**, 37 (1972).
1a. W. J. Price, *Analytical Atomic Absorption Spectrometry*, Heyden and Son Ltd., London, 1972.
2. L. E. Ranweiler and J. L. Moyers, *Environ. Sci. Technol.* **8**, 152 (1974).
3. C. E. Mulford, *At. Absorpt. Newsl.* **5**, 88 (1966).
4. J. D. Kinrade and J. C. VanLoon, *Anal. Chem.* **46**, 1894 (1974).
5. J. Dingman, S. Siggin, C. Barton and K. B. Hiscock, *Anal. Chem.* **44**, 1351 (1972).
6. A. A. EL-Awady, R. B. Miller and M. J. Carter, *Anal. Chem.* **48**, 110 (1976).
7. P. D. Goulden and P. Brooksbank, *Anal. Chem.* **46**, 1431 (1974).
8. S. Greenfield, H. McD. McGeachin and P. B. Smith, *Talanta* **22**, 1; **22**, 553 (1975).
9. V. A. Fassel and R. N. Kniseley, *Anal. Chem.* **46**, 110A; 1155A (1974).
10. R. F. Browner, *Analyst* **99**, 617 (1974).
11. H. Matsui, *Anal. Chim. Acta* **66**, 143 (1973).
12. H. Matsui, *Anal. Chim. Acta* **69**, 216 (1974).
13. J. Korkisch and D. Dimitriadis, *Talanta* **20**, 1287 (1973).
14. T. Yotsuyanagi, Y. Takeda, R. Yamashita and K. Aomura, *Anal. Chim. Acta* **67**, 297 (1973).
15. H. Weisz, *Microanalysis by the Ring-Oven Technique*, 2nd edn, International Series of Monographs in Analytical Chemistry, Vol. 37, Pergamon Press, Oxford, England, 1970.
16. T. R. Gilbert and D. N. Hume, *Anal. Chim. Acta* **65**, 451 (1973).
17. H. P. Yule, *Anal. Chem.* **37**, 129 (1965).
18. W. H. Zoller and G. E. Gordon, *Anal. Chem.* **42**, 257 (1970).
19. K. K. S. Pillay and C. C. Thomas, *J. Radioanal. Chem.* **7**, 107 (1971).
20. T. Fujinaga *et al.*, *J. Radioanal. Chem.* **13**, 301 (1973).
21. J. C. Russ (Ed.), *Energy Dispersion X-Ray Analysis*, ASTM Special Technical Publication No. 485, 1971.
22. D. E. Porter and R. Woldseth, *Anal. Chem.* **45**, 604A (1973).
23. C. F. Gamage, *X-Ray Spectrom.* **1**, 99 (1972).
24. J. V. Gilfrich, P. G. Burkhalter and L. S. Birks, *Anal. Chem.* **45**, 2002 (1973).
25. D. C. Camp, A. L. VanLehn, J. R. Rhodes and A. H. Pradzynski, *X-Ray Spectrom.* **4**, 123 (1975).
26. D. T. Carlton and J. C. Russ, *X-Ray Spectrom.* **5**, 172 (1976).
27. H. Watanabe, S. Berman and D. S. Russell, *Talanta* **19**, 1363 (1972).
28. Industrial Method No. 109-71W, Technicon Instrument Co. Tarrytown, N.Y.
29. Industrial Method No. 162-71W, Technicon Instrument Co. Tarrytown, N.Y.
30. Industrial Method No. 202-72W, Technicon Instrument Co. Tarrytown, N.Y.
31. A. Henriksen, *Automation in Analytical Chemistry*. Technicon Symposia, 1966, Vol. 1, Mediad, New York, 1967, p. 568.
32. *Standard Methods for the Analysis of Water and Wastewater*, 14th edn, APHA-AWWA-WPCF, 1975.
33. N. Rothstein and H. Zeitlin, *Anal. Lett.* **9**, 461 (1976).
34. G. C. Kunselman and E. A. Huff, *At. Absorpt. Newsl.* **15**, 29 (1976).
35. R. D. Wauchope, *At. Absorpt. Newsl.* **15**, 64 (1976).
36. J. Aggett and A. C. Aspell, *Analyst* **101**, 341 (1976).
37. R. C. Rooney, *Analyst* **101**, 678 (1976).
38. C. H. Kim, P. W. Alexander and L. E. Smythe, *Talanta* **23**, 229 (1976).

39. P. N. Vijan and G. R. Wood, *Analyst* **101**, 966 (1976).
40. P. N. Vijan and G. R. Wood, *At. Absorpt. Newsl.* **13**, 33 (1974).
41. M. McDaniel, A. D. Shendrikar, K. D. Reiszner and P. W. West, *Anal. Chem.* **48**, 2240 (1976).
42. E. L. Henn, *Anal. Chem.* **47**, 428 (1975).
43. K. C. Thompson, *Analyst* **100**, 307 (1975).
44. K. C. Thompson and R. C. Godden, *Analyst* **100**, 544 (1975).
45. M. G. Haywood and J. P. Riley, *Anal. Chim. Acta* **85**, 219 (1976).
46. S. S. Sandhu, *Analyst* **101**, 856 (1976).
47. M. K. John, H. H. Chuah and J. H. Neufeld, *Anal. Lett.* **8**, 559 (1975).
48. W. Davison and E. Rigg, *Analyst* **101**, 634 (1976).
49. J. Minczewski, J. Chwastowska and P. HongMai, *Analyst* **100**, 708 (1975).
50. S. A. Abbasi, *Anal. Chem.* **48**, 714 (1976).
51. D. Jagner and L. Kryger, *Anal. Chim. Acta* **80**, 255 (1975).
52. J. Gardiner and M. J. Stiff, *Water Res.* **9**, 517 (1975).
53. B. Salbu, E. Steinnes and A. C. Pappas, *Anal. Chem.* **47**, 1011 (1975).
54. E. S. Gladney and J. W. Owens, *Anal. Chem.* **48**, 2220 (1976).
55. K. H. Lieser and V. Neitzert, *J. Radioanal. Chem.* **31**, 397 (1976).
56. K. Kuga and K. Tsuji, *Anal. Chim. Acta* **81**, 305 (1976).
57. J. F. Elder, S. K. Perry and F. P. Bradley, *Environ. Sci. Technol.* **9**, 1039 (1975).
58. B. Holynska and K. Bisiniek, *J. Radioanal. Chem.* **31**, 159 (1976).
59. H. Agemian and A. S. Y. Chau, *Anal. Chim. Acta* **80**, 61 (1975).
60. P. N. Vijan, A. C. Rayner, D. Sturgis and G. R. Wood, *Anal. Chim. Acta* **82**, 329 (1976).
61. T. J. Forehand, A. E. Dupuy and H. Tai, *Anal. Chem.* **48**, 999 (1976).
62. J. W. Owens and E. S. Gladney, *At. Absorpt. Newsl.* **14**, 76 (1975).
63. T. H. Miller and W. H. Edwards, *At. Absorpt. Newsl.* **15**, 75 (1976).
64. M. J. Dudas, *At. Absorpt. Newsl.* **13**, 5 (1974).
65. A. H. C. Roberts, M. A. Turner and J. K. Syers, *Analyst* **101**, 574 (1976).
66. K. V. Krishnamurty, E. Shpirt and M. H. Reddy, *At. Absorpt. Newsl.* **15**, 68 (1976).
67. Y. Kuwae and T. Hasegawa, *Anal. Chim. Acta* **84**, 185 (1976).
68. P. Sutcliffe, *Analyst* **101**, 949 (1976).
69. E. P. Welsch and T. T. Chao, *Anal. Chim. Acta* **82**, 337 (1976).
70. K. Govindaraju, G. Mevelle and C. Chovard, *Anal. Chim. Acta* **48**, 1325 (1976).
71. F. D. Pierce and H. R. Brown, *Anal. Chem.* **48**, 670 (1976).
72. J. E. Caupeil, P. W. Hendrikse and J. S. Bongers, *Anal. Chim. Acta* **81**, 53 (1976).
73. F. W. E. Strelow, R. G. Bohmer and C. H. S. W. Weinert, *Anal. Chem.* **48**, 1550 (1976).
74. T. C. Z. Jaymon, S. Sivasubramanium and M. A. Wijedasa, *Analyst* **100**, 716 (1975).
75. T. Fukasawa and T. Yamane, *Anal. Chem.* **88**, 147 (1977).
76. R. A. Nadkarni and G. H. Morrison, *Anal. Chem.* **47**, 2285 (1975).
77. L. C. Bate, S. E. Lindberg and A. W. Andrea, *J. Radioanal. Chem.* **32**, 125 (1976).
78. E. Steinnes, *Anal. Chem.* **48**, 1440 (1976).
79. O. J. Kronborg and E. Steinnes, *Analyst* **100**, 835 (1975).
80. R. D. Giauque, R. B. Garrett and L. Y. Goda, *Anal. Chem.* **49**, 62 (1977).
81. P. V. Kulkarni and I. L. Preiss, *J. Radioanal. Chem.* **24**, 423 (1975).
82. R. Baum, W. J. Gutknecht, R. D. Willis and R. L. Walker, *Anal. Chim. Acta* **85**, 323 (1976).
83. B. C. Begnoche and T. H. Risby, *Anal. Chem.* **47**, 1040 (1975).
84. R. J. Bettger, A. C. Ficklin and T. F. Rees, *At. Absorpt. Newsl.* **14**, 124 (1975).

85. P. N. Vijan and C. Y. Chau, *Anal. Chem.* **48**, 1788 (1976).
86. A. Sugimae, *Int. J. Environ. Anal. Chem.* **34**, 185 (1976).
87. A. Albini, A. Cesana and M. Terrani, *J. Radioanal. Chem.* **34**, 185 (1976).
88. J. J. Paciga and R. E. Jervis, *Environ. Sci. Technol.* **10**, 1124 (1976).
89. M. Janssens, B. Desmet, R. Dams and J. Hoste, *J. Radioanal. Chem.* **26**, 305 (1975).
90. A. Owyla, M. Kasrai and R. Massoumi, *J. Radioanal. Chem.* **34**, 381 (1976).
91. T. G. Dzubay and R. K. Stevens, *Environ. Sci. Technol.* **9**, 663 (1975).
92. E. Brunix and E. VanMeyl, *Anal. Chim. Acta* **80**, 85 (1975).
93. R. C. Rooney, *Analyst* **100**, 465 (1975).
94. J. R. McDermott and I. Whitehill, *Anal. Chim. Acta* **85**, 195 (1976).
95. H. Freeman, J. F. Uthe and B. Flemming, *At. Absorpt. Newsl.* **15**, 49 (1976).
96. A. J. Thompson and P. A. Thoresby, *Analyst* **102**, 9 (1977).
97. K. Julsham and D. R. Braekkan, *At. Absorpt. Newsl.* **14**, 49 (1975).
98. T. J. Ganje and A. L. Page, *At. Absorpt. Newsl.* **13**, 131 (1974).
99. W. J. Simmons, *Anal. Chem.* **47**, 2015 (1975).
100. K. Matasunaga, T. Ishida and T. Oda, *Anal. Chem.* **48**, 1421 (1976).
101. G. B. Belling and G. B. Jones, *Anal. Chim. Acta* **80**, 279 (1975).
102. D. A. Shearer, R. D. Cloutier and M. Hidirohlou, *J. Assoc. Off. Anal. Chem.* **60**, 155 (1977).
103. M. Ihnat, *Anal. Chim. Acta* **82**, 293 (1976).
104. D. D. Siemer and L. Hagemann, *Anal. Lett.* **8**, 323 (1975).
105. M. R. Church and W. H. Robinson, *Int. J. Environ. Anal. Chem.* **3**, 323 (1974).
106. P. N. Vijan and F. R. Wood, *Talanta* **23**, 89 (1976).
107. A. Sugimae, *Anal. Chem.* **47**, 1840 (1975).
108. G. Forsberg, J. W. O'Laughlin and R. H. Megargle, *Anal. Chem.* **47**, 1586 (1975).
109. R. W. Andrews and D. C. Johnson, *Anal. Chem.* **48**, 1056 (1976).
110. M. W. Blades, J. A. Dalziel and C. M. Elson, *J. Assoc. Off. Anal. Chem.* **59**, 1234 (1976).
111. P. Lievens, R. Cornelis and J. Hoste, *Anal. Chim. Acta* **80**, 97 (1975).
112. A. Gaudry, B. Maziere, D. Comar and D. Nau, *J. Radioanal. Chem.* **29**, 77 (1976).
113. G. Guzzi, R. Pietra and E. Sabbioni, *J. Radioanal. Chem.* **34**, 35 (1976).
114. M. Gallorini, M. DiCasa, R. Stella, N. Genova and E. Orvini, *J. Radioanal. Chem.* **32**, 17 (1976).
115. B. C. Haldar and B. M. Tejam, *J. Radioanal. Chem.* **33**, 23 (1976).
116. K. Norrish and J. T. Hutton, *X-Ray Spectrom.* **6**, 6 (1977).
117. J. T. Hutton and K. Norrish, *X-Ray Spectrom.* **6**, 12 (1977).
118. K. I. Strausz, J. T. Purdham and O. P. Strausz, *Anal. Chem.* **47**, 2032 (1975).
119. G. S. Kirkbright and M. Sargent, *Atomic Absorption and Fluorescence Spectroscopy*, Academic Press, London, 1974.

CHAPTER 3

CLASSIFICATION OF METAL COMPOUND SPECIES

Most of the methods described in Chapter 2 determine the elemental composition of the sample. In some cases however the chemical form of the metallic element is of concern, particularly in the case of metals which form volatile alkyl compounds. These materials are sufficiently volatile to exist in the atmosphere as gases or vapours and their toxicity makes them a potential health hazard. These same materials may be formed in natural waters by the alkylation of inorganic compounds by microorganisms. The metals which are of particular concern are those which have been shown to be methylated in nature, namely mercury, lead, selenium and arsenic.

With metal ions in solution it is often desired to determine their particular ionic form, i.e. whether they are 'free' ions, whether they are complexed, or whether they are adsorbed on to particulate matter. This is important in that it determines their effect on the plant and animal life in the water and affects their transportation or immobilization.

METAL ALKYLS

Mercury

There has been a considerable amount of work carried out to distinguish methylmercury from other mercury forms in environmental samples. This work has been concerned to a great extent with the examination of biological samples, particularly fish and shellfish. This is because of the recognition of the toxicological effect of methylmercury and, in particular, the demonstration that the cause of Minamata disease is the attack on the central nervous system by the methylmercury radical, (CH_3Hg^+). In 1958 a number of cases of a strange neurological disease were reported in Japan, in the area around Minamata Bay. Out of 121 cases reported, 46 resulted in death, and there was severe permanent disability among the survivors. The disease was correlated to the eating of fish and shellfish from Minamata Bay, but it was not until 1964 that the cause was established as methylmercury. There was a further occurrence of the Minamata disease at Nilgata, Japan in 1965. At that time the source of methylmercury was thought to be the discharge of industrial waste containing methylmercury into the water. In the early 1960s there became an awareness of a mercury problem in Sweden, with the finding of high levels of mercury in seed-eating birds. This was believed to be due to their eating seeds which had been mercury-treated before planting. There were also high levels of mercury found in fish. In 1966 Westoo[1] showed the presence of methylmercury in fish, in fact the major portion of the mercury present in fish tissue was methylmercury. Jensen and Jernelov[2] showed that there was a widespread distribution of microorganisms that could transform inorganic and other organomercurials to methylmercury. The same sort of history of recognition of a mercury problem, first through seed-eating birds and then through fish, occurred in Canada in the late 1960s. Since this recognition of the serious toxicological effects of mercury discharges to the environment there has been developed a whole array of techniques for measuring, first the level of 'total' mercury and then the methylmercury in environmental samples. The majority of methods used for the determination of 'total' mercury use atomic absorption spectroscopy as the measurement technique. The most common form is the cold vapour process and this is described in Chapter 2.

Tissues

The methods for determining methylmercury in tissue samples for the most part involve an extraction of the methylmercury, then a clean-up procedure with separation and quantification by gas–liquid chromatography (g.l.c.). Most extraction processes are based on the work of Gage[3] who found that organomercurials, such as methylmercury and phenylmercury, can be extracted from tissue into benzene in the presence of high concentrations of hydrochloric acid. Inorganic mercury does not extract into the benzene under these conditions.

The procedure for the analysis of fish tissues described by Bouveng[4] was as follows. Methylmercury was extracted from the homogenized tissue with toluene in the presence of copper sulfate, sodium bromide and sulfuric acid. The methylmercuric bromide was back-washed from the toluene with sodium thiosulfate solution, potassium iodide was added, and the resulting methylmercuric iodide was extracted into benzene and analysed by gas chromatography. The gas chromatographic analysis was carried out on 7% Carbowax 20M on Varaport-30 (100–200 mesh) at 140 °C. The carrier gas was nitrogen and an electron capture detector was used. The sensitivity was reported as 10^{-11} g and samples containing 0.5 μg kg^{-1} could be analysed. The recovery of added methylmercury was reported to be approximately 90%.

A modified version of this procedure was described by Uthe et al.[5] The extraction of the methylmercuric bromide into toluene was carried out by shaking the sample and solvents in a capped centrifuge tube containing two glass marbles. The layers were separated by centrifuging the tube and 5 ml of the toluene layer were extracted twice with an ethanolic solution of sodium thiosulfate. (The use of ethanol prevented the formation of emulsions.) The methylmercury thiosulfate complex was converted to the iodide by the addition of potassium iodide. The methylmercuric iodide was then extracted into benzene for analysis by gas chromatography. The chromatographic analysis was carried out at 170 °C on 7% Carbowax 20M on high-performance Chromosorb W (80–100 mesh) using nitrogen as the carrier gas. Recoveries of methylmercury added to the sample before the grinding step were 99 ± 5%. Standard curves were determined using methylmercuric chloride rather than methylmercuric iodide because of the lability of methylmercuric iodide. The sensitivity of the chromatographic system was such that 10^{-11} g of methylmercuric chloride could be detected. There were still some extracts which required further clean-up and this was carried out by passing the final benzene extract through a small column of Florisil overlaid with sodium sulfate.

There are some methylmercuric compounds which do not liberate methylmercury for extraction by the action of strong halide acids alone. Bis-dimethylmercury sulfide is one example and the effectiveness of certain metal ions in breaking down these compounds was investigated by Fujiki.[6] He found that mercuric chloride, cuprous chloride and silver chloride were effective (sodium, calcium, magnesium, zinc and ferrous chlorides were not) and cuprous chloride was the one recommended. If mercuric chloride is added to the sample before extraction there is the risk that any dimethylmercury will react with the mercuric ion to produce monomethylmercury.

A method for the determination of both methylmercury and dimethylmercury is reported by Hartung.[7] The sample was homogenized with water containing sodium tetraborate and cysteine hydrochloride. This was then extracted three times with toluene to extract the dimethylmercury. The aqueous phase was then acidified with an equal volume of concentrated hydrochloric acid and extracted three more times with toluene to extract the methylmercury. The toluene layer containing the dimethylmercury was saponified for two hours with potassium hydroxide, and was then washed twice with water and dried over sodium sulfate. The dimethylmercury was next converted to methylmercuric bromide by refluxing with an aqueous solution of mercuric and potassium bromide. The toluene layer was then washed, dried and analysed by gas chromatography. The toluene from the first stage of the separation, which contained methylmercuric chloride, was extracted with fresh cysteine borate solution. The solution was then acidified with hydrochloric acid and the methylmercuric chloride was analysed by gas chromatography. These analyses were carried out at 100 °C on 11% QF1 + OV17 on Gas-chrom Q (80–100 mesh) using nitrogen as a carrier gas.

The levels of mercury found in fish are of the order of fractions of a microgram of mercury per gram of wet fish. (The level of mercury at which fish are declared unfit to eat is 0.5 μg g^{-1} in Canada.) The greater fraction of this mercury, about 90%, is commonly present as methylmercury.

A method for estimating the methylmercury content of cereal products and fish was described by Newsome.[8] For the analysis of cereal products 10 g of the sample was ground for 5 min with a mixture of benzene and 90% formic acid. The homogenate was filtered and a portion of the filtrate was layered onto a silicic acid column. Approximately 1 ml min^{-1} of benzene was passed through the column and the fraction eluting between 15 and 55 ml was collected. This fraction was shaken with cysteine acetate solution and, after separation, a portion of the cysteine acetate solution was acidified with 48% hydrobromic acid. The resulting methylmercuric bromide was extracted into benzene for gas chromatographic analysis. Recoveries of added methylmercury reported were 99 $\pm$ 6% for oats and 105 $\pm$ 11% for wheat flour.

A large part of the analytical work on organomercurials has been concerned with the determination of methylmercury because of its demonstrated toxicity and the fact that it is synthesized from other mercury forms in nature. Other organomercurials may be found in biological samples since they are, or have been, used as fungicides. These fungicides have been applied to seeds, to prevent rot, and to fruit crops, such as apples, to prevent scab. There are three main classes of organomercurials which are of interest. These are:

1. Alkylmercury compounds: methylmercury, Me—Hg—X
   ethylmercury, Et—Hg—X
2. Alkoxyalkyl mercury compounds: methoxyethylmercury,
   MeO—Et—Hg—X
   ethoxyethylmercury, EtO—Et—Hg—X
3. Arylmercury compounds: phenylmercury, Phenyl—Hg—X
   tolylmercury, Tolyl—Hg—X

where X is an anion such as sulfate, nitrate or chloride.

A procedure for the extraction of these materials from apples, tomatoes and potatoes, and for their subsequent identification and determination by thin-layer and gas–liquid chromatography is described by Tatton and Wagstaffe.[9] The procedure used for the separation of methylmercury is one not suitable for the extraction of the alkoxyalkylmercurials since they readily decompose in acid solutions. In the system described, 5 g of the sample (chopped apple or potato peel or macerated tomato) were ground with a mixture of 10 ml of propan-2-ol and 5 ml of alkaline cysteine hydrochloride solution (1% aqueous solution adjusted to pH 8 with ammonia). After allowing the liquor to settle the clean layer was decanted and the extraction repeated twice more with further portions of the extractant solution. The combined extracts were then centrifuged. The clear liquid was separated, diluted with 700 ml of 4% sodium sulfate and the solution washed with three 50 ml portions of diethyl ether. The organomercurials

were extracted from the aqueous solution using three 25 ml portions of a
0.005% solution of dithizone in diethyl ether. The combined extracts were
dried, by passing them through a short column of granular anhydrous sodium
sulfate, and concentrated to a suitable volume, usually 5 ml, in a Kunderna-
Danish evaporator. The final solution was then examined by thin-layer chromatography (t.l.c.) or by g.l.c.

In this extraction procedure the anion (X) that was originally associated with
the organomercurial is replaced with dithizone. This (as discussed by Tatton and
Wagstaffe[9]) is a necessary procedure since anions such as sulfate, nitrate,
acetate tend to make the compound ionic and water soluble, while anions such
as the halogens or dicyandiamide tend to make the compound non-polar and
soluble in organic solvents. Hence if the original anion is retained the separations are dependent on the particular anion rather than on the type of organomercurial radical.

A number of thin-layer separation systems were examined. Using silica gel and
alumina as absorbants, various mixtures of hexane–acetone and light petroleum–
acetone were used as the mobile phase. Separations of the pure compounds for
use as standards were made using a t.l.c. plate with a thick layer of stationary
phase. The gas-chromatographic separations were made at 180 °C on 2%
polyethylene glycol succinate on Chromosorb G (60–80 mesh) in a glass column,
1.5 m long 3 mm i.d. using nitrogen as the carrier gas. If arylmercury compounds
were present a separation was made on 1% polyethylene glycol succinate on
Chromosorb G in a column 1.2 m long. The recoveries obtained were 85–95%
for samples spiked with 1.0, 0.1 and 0.01 $\mu g\,l^{-1}$ of methyl-, ethyl- and ethoxy-
ethylmercuric chlorides and 5 and 0.5 $\mu g\,g^{-1}$ of phenyl- and tolylmercuric
acetates.

Sediments

In applying the same sort of extraction techniques as used for fish tissue to
sediments, Westoo[10] found that there were problems because of the presence of
interfering non-volatile sulfur compounds of methylmercury. She found it
necessary to pretreat the sediments with mercuric chloride to decompose these
sulfur compounds in order to give good recoveries of methylmercuric chloride.
This is a similar problem to that found in fish tissue investigated by Fujiki (see
p. 45). Nishi[11] studied the removal of these interfering sulfur compounds and
found that mercuric chloride was the most effective reagent but that cuprous or
cupric salts could also be used. Hence the procedure described by Bouveng[4] and
Uthe[5] in which copper sulfate and sodium bromide are added, should also be
applicable to sediments and it has been reported[12] that Uthe's procedure does
give good recovery of methylmercuric chloride from sediments. If mercuric
chloride is used to decompose the sulfur compounds, the excess must be precipitated before the extraction, otherwise some of it may extract and interfere in
the gas chromatographic analysis.

The determination of methylmercury in estuarine sediments has been reported by Andren and Harriss.[13] They determined the methylmercury, and the total mercury in sediments from Mobile Bay, the Mississippi Delta and the Florida Everglades. They determined methylmercury as the total dimethylmercury and monomethylmercury, by conversion to the bromide and analysis by gas chromatography. The total mercury was determined by atomic fluorescence. The results they obtained are shown in Table 3.1. It is seen that the fraction of the mercury

TABLE 3.1
Methylmercury and total mercury in some surface sediments (from Andren and Harriss, Ref. 13)

Sample	Total Hg/ng g^{-1}	MeHg/ng g^{-1}	MeHg as % age of total Hg
Mobile Bay			
1	210	0.06	0.03
2	600	0.19	0.03
Mississippi River			
1	140	<0.02	<0.01
2	80	<0.02	<0.01
3	510	0.05	0.01
4	570	0.05	0.01
Everglades			
1	120	0.06	0.05
2	120	0.08	0.07
3	490	0.12	0.03

that is present as methylmercury is very small; it averages 0.03% and the maximum value is 0.07%. They studied the methylmercury fraction as a function of sediment depth and found that the fraction decreased with increasing distance from the surface.

A 'low cost' method for the determination of monomethylmercury in aquatic samples, i.e. water, biomass or inorganic and organic sediments, is described by Bisogni and Lawrence.[14] In their procedure the organic mercury was extracted and converted to elemental mercury for measurement by the cold vapour atomic absorption technique. The sample was acidified with hydrochloric acid and then extracted with benzene. The benzene extract was shaken with a buffered cysteine acetate solution to transfer the mercury to the aqueous phase. This aqueous solution was then acidified and treated with potassium permanganate and potassium persulfate at 95 °C for 2 h to oxidize organic compounds. The solution was then sparged with nitrogen to remove any residual benzene or any chlorine formed in the oxidation step. Hydroxylamine hydrochloride and stannous chloride were added and the elemental mercury which was produced was swept to an absorption cell in a photometer for measurement. In this work it was found that the addition of sodium chloride minimized the extraction of inorganic mercury into the benzene phase, presumably through the formation of such complexes as $NaHgCl_3$ or Na_2HgCl_4. Other organic mercury compounds would extract into the benzene phase and be measured; for example, it was shown that phenylmercury compounds would extract. The basis of the claim for the measurement being one of monomethylmercury was that this is usually the only organic form of mercury found in aquatic samples. It was proposed that the organic mercury extracted be examined by thin-layer chromatography and if significant quantities of other organomercurials were found the sample should be quantified using gas chromatography.

A similar approach to the determination of organomercurials in the analysis of sediments was taken by Matsunaga and Takahashi.[15] In their procedure 10–20 g of sediment were taken and 10 ml of 2 M hydrochloric acid added. The flask was then allowed to stand for two days. The sample was filtered through a glass filter and the residue washed with hydrochloric acid. The filtrate was shaken with 40 ml benzene to extract the organomercurials. Back extraction into an aqueous phase was made with 20 ml of ammoniacal solution of glutathione. To a gas washing bottle was added 150 ml of water, 10 ml of 10 M sodium hydroxide, 2 ml of 1000 ppm copper solution and 5 ml of 5% stannous chloride dihydrate solution. This solution was swept with nitrogen to remove any mercury in the reagent solution and the aqueous back-extract from the sample was then added. The elemental mercury vapour produced was concentrated by passing the nitrogen purge through a tube packed with gold granules. This mercury was then released by heating the gold granules to 500 °C. Measurement was made in a quartz cell in an atomic absorption spectrophotometer. As has already been discussed this method measures the total organic mercury level.

The system used for the reduction to elemental mercury is based on the work of Umezaki and Iwamoto[16] who found that organomercurials can be reduced directly to elemental mercury by reaction with stannous chloride in alkaline solution in the presence of copper. In this way the oxidation step to convert organomercurials to inorganic mercury, that is necessary in the 'conventional' acid reduction process, is not needed. The only source of interference in this method is inorganic mercury that may be present in the ammoniacal glutathione solution. This is overcome in the method by allowing the ammoniacal glutathione solution to stand in a glass container for about a week. Any inorganic mercury then adsorbs onto the glass walls; the halflife of the mercury in solution was found to be about 2 days. The organic mercury found in two sediment samples from unpolluted locations was 0.22 ± 0.03 and 0.43 ± 0.03 ng mercury per gram sediment (dry weight).

Water samples

A method for the determination of methylmercury in lake waters has been described by Chau and Saitoh.[17] Their procedure was to filter the water immediately after collection, acidify it with sulfuric acid and then extract 5 litres of sample with 300 ml benzene for 5–6 h. The extraction procedure was based on the method developed by Nishi and Horimoto which uses a continuous extractor. The benzene was then shaken with four 5 ml aliquots of a 0.2% L-cysteine solution. The combined L-cysteine extracts were then combined and adjusted to 0.6 N with hydrochloric acid. This acidified solution was extracted with four 5 ml aliquots of benzene. The four aliquots were combined and concentrated to 2 ml under reduced pressure. Aliquots of this benzene extract were injected into the gas chromatograph for determination. The separations were made in a column six feet long packed with 5% diethylene glycol succinate on Chromosorb W, 60–80 mesh, acid washed and pre-coated with sodium chloride. The oven temperature was 130 °C and the carrier gas was nitrogen.

In this work recoveries were determined by using ^{203}Hg labelled methylmercuric chloride. It was found that after 2 h of extraction in the continuous extractor, recovery of added methylmercuric chloride was 45–50%. After 5–6 h the recovery was 96–98%. The overall recovery in all the operations was about 88%. It was shown that methyl-, ethyl- and phenylmercuric chloride would be determined by the method.

The technique was used on samples from several lakes in Canada: Lake St Clair, St Clair River, Saskatchewan River, Pinchy Lake, B.C. and Clay Lake, Kenora where very high mercury levels in fish have been previously reported. Methylmercury was found only in the following samples: Lake St Clair—0.6 ng l^{-1}; Clay Lake, Kenora—0.5 ng l^{-1} (central lake), 1.0 ng l^{-1} (near shore); Pinchy Lake, B.C.—1.7 ng l^{-1}. No ethyl- or phenylmercury was found. No methylmercury was found in any of the four Great Lakes.

Air samples

In air sampling the concern has been mainly to distinguish the alkylmercury compounds from elemental mercury since they are both volatile. Kimura and Miller[18] in a study of organomercurial fungicide decomposition in soil were able to trap ethylmercury and methylmercury vapours separately from metallic mercury. Two absorbers were used, the first one containing a solution of sodium carbonate and disodium phosphate, and the second an acid solution of potassium permanganate. When the air was drawn through the absorbers the methyl- and ethylmercury were absorbed in the carbonate–phosphate absorber, the mercury vapour passing through and being absorbed in the acid permanganate. The absorbed mercury was determined with dithizone. Alkylmercury compounds in the range of a nanogram per cubic meter of air were determined in this way.

Christie et al.[19] describe a method for the absorption of organomercurial vapours from the air in the range of 10 μg per cubic metre of air. They used a glass-fibre pad treated with cadmium sulfide to absorb ethylmercuric chloride, ethylmercuric phosphate, diphenylmercury and methylmercuric dicyandiamide. A fluidized bed of active carbon was also used to absorb the same mercurials; it would in addition absorb diethylmercury. A scheme was proposed for distinguishing between the alkyl mercurys, metallic mercury vapour and mercury-bearing dust. This scheme used the glass-fibre pad treated with cadmium sulfide, the activated carbon and also iodized carbon. Measurements were made by heating the absorbent containing the mercury compounds and comparing the colours produced on selenium sulfide test papers.

Langley[20] studied the production of dimethylmercury in work on the methylation of mercury under laboratory conditions. He used sediments taken from a river below a chlor-alkali plant outfall. The dimethylmercury produced was trapped by passing it into a gas-washing bottle containing acid mercuric nitrate. This converted the dimethylmercury to monomethylmercury the level of which was then determined.

Confirmatory analysis

The most common way in which the organomercurial species have been identified and quantified is by gas chromatography using an electron capture detector. The high sensitivity of this analytical method results from the high absorbing power of mercury in this detector. However the positive identification of a specific material requires more than that its retention time in the chromatographic procedure is the same as that of the standard material. In addition the samples that are injected into the gas chromatograph should contain as little foreign material as possible if a clean separation is to be made. This need for a rigorous clean-up of the samples is why the multi-stage extraction procedures between solvent and water and back again are carried out in the organo-mercurial analytical procedures.

The confirmatory procedures which were used, e.g. by Westoo[1] in the original analysis of methylmercury in fish, involved both gas chromatography and thin-layer chromatography. The benzene extracts of fish tissue gave spots with the same R_F as methylmercuric chloride. These spots, when removed from the plate, extracted and injected into the gas chromatograph, gave a peak with the same retention time as authentic methylmercuric chloride. The fish extracts and methylmercuric chloride were treated to form the dithizonate, bromide, iodide or cyanide. These materials were then chromatographed and in all cases the fish extract and the methylmercury behaved in the same manner. Later, mass spectroscopy was used to confirm the presence of methylmercury in the benzene extracts of fish tissue.

Jensen[21] has suggested that the benzene solution of fish extract which has given a peak for methylmercury in the gas chromatograph should be shaken with aqueous silver sulfate and then rechromatographed. Any methylmercuric chloride would be converted to methylmercuric sulfate which would then partition into the aqueous phase. Hence if the original peak were methylmercuric chloride it should not be present in the second chromatogram. Other derivatives of the methylmercury can also be prepared, such as methylmercury methylene bromide which has a different retention time to that of methyl-mercury. With the increasing availability of gas chromatograph–mass spectrometer coupled systems, confirmation of the chromatogram peaks as methyl-mercury can be made by the mass spectrometer.

An alternative to the electron capture detector is to use a mercury-specific device. Bache and Lisk[22] passed the effluent from the gas chromatograph through a helium plasma and measured the mercury using the 253.7 nm emission line. Longbottom[23] passed the gas chromatograph effluent through a flame ionization detector and determined the mercury in the combustion products by measuring the absorption at 253.7 nm in a cold-vapour cell.

A review of the methods for the determination of mercury and organo-mercury compounds in environmental materials has been published by Uthe and Armstrong.[12]

Lead

Tetramethyl lead, tetraethyl lead and the mixed alkyl derivatives are used as 'anti-knock' compounds in gasoline. Although the major part of these materials is converted to inorganic lead when the gasoline is burned, evaporation of gasoline is probably the major source of lead alkyls entering the environment. It has also been shown by Wong et al.[24] that lead can be methylated by micro-organisms that occur in lake water. In this work it was found that Me_3Pb salts are most readily converted to Me_4Pb in the sediments but that Me_4Pb could also be generated from some inorganic lead salts such as the chloride or nitrate.

In air sampling a distinction can be made between volatile lead and particulate lead; the volatile lead presumably comprises organolead compounds. In the method described by Snyder[25] for the 'determination of trace amounts of organic lead in air', filtered air was passed through activated carbon, the adsorbed lead was then dissolved from the carbon and determined colorimetrically. The adsorber was made from a glass tube of 24 mm i.d. in which was placed 10 g of activated carbon (30–50 mesh) on a glass wool support. The filtered air was passed through the adsorber at a rate of about 20 l min^{-1} until approximately 200 m^3 of air was sampled (this took about a week). At the end of the sampling period the activated carbon was heated with aqua regia at 100 °C overnight and then heated with further nitric acid solution. The carbon was then separated by decantation and filtering. To the extract was added a nitric–perchloric acid mixture and the solution was evaporated until fumes of perchloric acid were obtained, to oxidize any remaining carbon. The solution was diluted into a 250 ml absorption cell and a buffer solution containing dibasic ammonium citrate and potassium cyanide was added, followed by a reagent solution containing ammonium citrate, potassium cyanide, sodium sulfite and ammonia. Dithizone solution was then added and the absorption at 510 nm measured. Sodium diethyldithiocarbamate (or EDTA) solution was added, and the absorbance read again. The diethyldithiocarbamate decomposes the lead dithizonate and the difference between the two absorbance readings represents the amount of lead present in the sample. Diethyldithiocarbamate decomposes the lead dithizonate quickly but it also decomposes the excess dithizone so that the second absorbance reading must be taken within 30–45 s of the addition. The EDTA solution decomposes the lead dithizonate more slowly, but the solution colour after this decomposition is more stable. The interferences in this method are of course particulate lead, which must be removed in the filtration step, and other metals which react with dithizone in ammoniacal cyanide solution to give coloured complexes. These are bismuth, thallium, tin and indium. However there are no known volatile compounds of these metals likely to be present in ambient air. As a confirmation that the colour measured was not due to metal dithizone complexes other than lead, the lead dithizonate from several air samplings was collected, concentrated and examined spectrographically. With this method organic lead can be measured down to 0.01 μg m^{-3}. Sampling carried out in Los Angeles over a period of six weeks showed an average organic lead content of 0.078 μg m^{-3} and an average particulate lead content of 3.56 μg m^{-3}.

The lead alkyls can be separated by gas–liquid chromatography. Parker et al.[26] separated the methyl–ethyl lead alkyls in gasoline anti-knock mixture, collected portions of the effluent gas from the chromatograph in methanolic iodine solution and measured the lead content of each fraction spectrophotometrically with dithizone. The alkyl lead compounds separated were Me_4Pb, Me_3EtPb, Me_2Et_2Pb, $MeEt_3Pb$ and Et_4Pb.

Bonelli and Hartmann[27] separated the same materials in gasoline by gas–liquid chromatography, using an electron capture detector. In this method it was necessary to use a scrubber containing silver nitrate to remove the halogenated scavengers (such as ethylene bromide) used in gasoline. Cantuti and Cartoni[28] measured the tetraethyl lead concentration in air by drawing the air sample through a small tube containing a gas chromatographic support and liquid phase. The tube was kept either at ambient temperature or at 0 °C with ice. After the sample was collected the tetraethyl lead was desorbed by heating the tube and the gas swept into the gas chromatograph. Detection was by electron capture. The separator column in the gas chromatograph, and the collection tube, were packed with Chromasorb P, 80–100 mesh, containing 10% silicone rubber SE-52. The column oven was at 80 °C and the carrier gas was nitrogen. With this method 0.1 ppm of tetraethyl lead could be detected in air.

The electron capture detector is sensitive to lead alkyl compounds, but it lacks specificity and in environmental samples it is desirable to use a detector that is element specific. A system in which a monochromator and photomultiplier were added to a regular flame ionization detector is described by Hill and Aue.[29] With this system the elements eluting from the gas chromatograph can be identified and quantified by monitoring one of their emission spectrum lines. It was found that, using a hydrogen-rich flame, it was possible to measure metallic elements such as iron, tin and lead in the lower nanogram range. The same authors[30] also describe a hydrogen-rich flame ionization detector sensitive to the same metals. Emission spectroscopy using a helium plasma[22] and atomic absorption with cold vapour detection[23] have been described for the gas chromatographic analysis of mercury compounds.

'Flameless' atomic absorption spectroscopy has been used as a detector for gas chromatographic separation. Segar[31] describes a system in which the effluent from a gas chromatograph was passed into the heated tube of a graphite furnace via a tungsten transfer line. The furnace operated at about 2700 °C. Separation of a mixture of the five methyl and ethyl lead alkyl mixtures in a gasoline sample was obtained; the sensitivity of the detection was about 10 ng of Pb.

The use of flame atomic spectroscopy for detection in the gas–liquid chromatographic separation of the tetraalkyl lead compounds in the atmosphere has been reported by Chau et al.[32] The alkyl lead compounds in the air sample were trapped by a system similar to that used by Cantuti and Cartoni.[28] A U-tube made of glass, 26 cm long 6 mm o.d. was packed with the material used in the gas chromatographic column, in this case 3% OV-1 on Chromosorb W, 80–100 mesh. The U-tube was immersed in a dry-ice bath at −80 °C. The air was drawn, first through a 0.45 μm Sartorius membrane filter, and then through the U-tube at 130–150 ml min^{-1} using a peristaltic pump. After a known volume of air had been passed through the trap the ends of the U-tube were closed for transport to the laboratory. The exposed U-tube could be stored in a dry-ice chest for at least four days without loss of any alkyl lead. For analysis the U-tube was placed in water at 50 °C and the vaporized sample was swept into the gas chromatograph. The gas chromatograph column was of glass 1.8 m long 6 mm diameter packed with 3% OV-1 on Chromosorb W, 80–100 mesh. The carrier gas was nitrogen. The oven was maintained at 40 °C for 2 min then programmed for a 5 °C rise per minute to 90 °C. The effluent from the gas chromatograph was led, via a 2 mm diameter stainless steel tube, to the nebulizer connection on the burner of the atomic absorption spectrophotometer. The spoiler in the cloud chamber was removed and a glass tube was installed in the cloud chamber to reduce any adsorption of the alkyl compounds on the Penton lining. An air–acetylene flame was used and the absorbance of the lead line at 217 nm was measured. The detection limit of the determination was about 0.08 μg of each lead alkyl compound, expressed as Pb.

In further development of the technique[33] the flame detection was replaced by a system using a small electrically heated silica tube placed with its axis in the light path of the instrument. In this system the gas enters the centre of a silica tube that is electrically heated to about 1000 °C. Hydrogen is mixed with the gas and this burns inside the tube to produce atomic lead. A similar system is used in the determination of selenium alkyls and is described further below. An increase in sensitivity over the regular flame of three orders of magnitude is obtained, and the detection limit of this method is about 0.1 ng Pb for each of the alkyl leads. The possible interferences with this technique are volatile organic materials, and this interference is overcome by the use of the background corrector.

Selenium

It has been shown that selenium can be methylated by a variety of fungi, plants and animals to produce the volatile dimethylselenide and dimethyldiselenide. Because of the potential health hazards associated with the volatile selenium compounds, there has been interest in establishing analytical methods for these materials in the atmosphere. A number of organic sulfur compounds have been identified in plant tissue by gas–liquid chromatography. Selenium forms many compounds analogous to those formed by sulfur and there has been interest in establishing the characteristics of the selenium analogues so that if they occur in plants they can be identified. Evans and Johnson[34] prepared seven alkyl selenium compounds and studied their separation characteristics in a gas–liquid chromatograph. The compounds studied were: dimethylselenide, diethylselenide, dipropylselenide, dimethyldiselenide, diethyldiselenide, dipropyldiselenide and ethylselenocyanate. In this work a flame ionization detector was used. A similar study was carried out by Benes and Prochazkova,[35] who prepared a number of organoselenium compounds and determined their separation characteristics relative to analogous sulfur compounds by gas–liquid chromatography. The selenium compounds studied were: dimethylselenide, diethylselenide, dipropylselenide, di-*n*-butylselenide, phenylselenol, phenylmethylselenide and phenylethylselenide. The detector used on the gas chromatograph in this case was an ^{90}Sr detector.

The volatile dimethylselenide and dimethyldiselenide have been determined by gas–liquid chromatography. In the case of a study by Barkes and Fleming[36] of their production from soil fungi the gas above the culture could be injected directly into the gas chromatograph.

In the work reported by Vlasakova *et al.*,[37] the selenium compounds expired by the organisms could not be measured directly. They were concentrated from the gas by adsorption onto an Alusil/sodium aluminosilicate granular absorbent or onto activated charcoal, in a similar manner to that used for the adsorption of lead alkyls. The adsorbed compounds were released by heating and swept into the gas chromatograph. Animals (e.g. rats) were fed with ^{75}Se and the expired air analysed. The chromatograph peaks for dimethylselenide and dimethyldiselenide were identified by the simultaneous collection of fractions of the effluent gas on a charcoal adsorbent and determination of their radioactivity. The separations were made on a column, using 15 % SE-30 silicone polymer or 10 % Apiezon L on 80–100 mesh Gas Chrom P carrier. Nitrogen was the carrier gas and a temperature programme from 20 to 240 °C was used on the oven. The detector was a flame ionization detector. It was noted in this work that not all of the ^{75}Se could be desorbed from the Alusil adsorbent used to collect the samples. About 5 % of the radioactivity remained even after heating to 500 °C.

The ratio of the selenium gases produced by the animals was 80% dimethyl-selenide: 20% dimethyldiselenide.

A method for the analysis of dimethylselenide and dimethyldiselenide is described by Chau *et al.*[38] This is similar to the method described above for the determination of the alkyl lead compounds. The technique was developed to analyse the biologically generated volatile selenium compounds in the atmosphere of lakewater sediment systems, but it can also be used to determine these compounds in atmospheric air samples. The sample was collected by passing the air through a U-tube trap (26 cm long, 6 mm diameter) filled with 3% OV-1 on Chromosorb W. The trap was immersed in a dry-ice bath at $-80\,°C$. After the sample was collected, the U-tube was heated to about 100 °C (most conveniently with a household toaster), and the volatilized selenium compounds were swept into the gas chromatograph. The separations were made on a column packed with the adsorbent above and a temperature programme from 40 to 120 °C was used on the oven. The carrier gas was nitrogen. The effluent gas was passed into

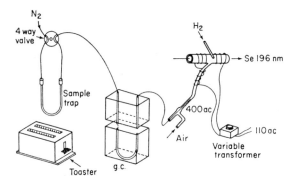

Fig. 3.1. Schematic diagram of GC-AAS set-up used for determination of methyl-selenides. (From Ref. 38.)

an electrically heated silica tube mounted in the light path of an atomic absorption spectrophotometer where atomic selenium was produced and measured at the 196 nm selenium line. The arrangement is shown schematically in Fig. 3.1; a typical recorder tracing obtained with dimethylselenide and dimethyldiselenide is shown in Fig. 3.2. It was found that 0.1 ng of selenium occurring as either of the alkyl selenides could be measured.

The silica tube was 6 cm long, 7 mm i.d. and heated to about 1000 °C. Hydrogen was fed into the tube with the gas from the gas chromatograph and burnt inside the tube. It was found that there was interference from the other materials, particularly chlorinated hydrocarbon, that may also be trapped with the volatile selenium compounds. These have a broad band absorbance in the spectral region of 196 nm and the absorbance in some cases was so high that the deuterium background corrector could not make sufficient compensation. To remove this interference a small pre-combustion furnace was installed in the line leading from the gas chromatograph, and air was added at this point to burn the contaminants. The furnace was a piece of silica tube 3 mm o.d. and 10 cm long that was electrically heated. When the contaminating solvents were combusted the background corrector could compensate for the combustion

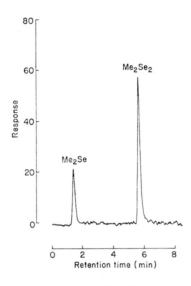

Fig. 3.2. Recorder tracing of dimethylselenide (10 ng Se) and dimethyldiselenide (16 ng Se). (From Ref. 38.)

products and any other non-selenium material in the sample. With this element-specific detection the response is only to selenium compounds so that the problem of resolution of the peaks from different materials such as sulfur compounds is not necessary.

Arsenic

It has been known for a long time that arsenic can be methylated by biological processes. In 1847 Gosio reported that a strong garlic-like odour was produced when a number of fungi were grown in the presence of sodium arsenite. Bignelli in 1900 erroneously concluded that this gas was diethylarsine. Challenger et al., in 1933, showed that the gas was trimethylarsine. McBride and Wolfe[39] showed that dimethylarsine was produced by certain methanogenic bacteria. The route for this conversion was shown to be:

$$
\begin{array}{cc}
\text{HO—As—OH} & \text{As—OH} \\
\|\ & \rightarrow\ \| \\
\text{O} & \text{O} \\
\text{arsenate} & \text{arsenite}
\end{array}
$$

$$
\begin{array}{ccc}
\text{CH}_3 & \text{CH}_3 & \text{CH}_3 \\
| & | & | \\
\text{HO—As—OH} \rightarrow & \text{HO—As—CH}_3 \rightarrow & \text{As—CH}_3 \\
\| & \| & | \\
\text{O} & \text{O} & \text{H} \\
\text{methylarsonic acid} & \text{dimethylarsinic acid} & \text{dimethylarsine}
\end{array}
$$

That the methylarsonic acid and the dimethylarsinic acids are in fact found in the environment has been shown by Braman and Foreback.[40] They examined a number of environmental samples and demonstrated that the arsenic was present in the four forms: arsenite, arsenate, methylarsonic acid and dimethylarsinic acid. The basis for the separation was the determination of conditions under which these materials could be reduced to the corresponding arsines, and selective volatilization of the arsines from a cold trap. In the procedure used it was found that As(III) was the only arsenic form reduced to arsine by sodium borohydride at pH 4–9. As(V) must first be reduced to As(III) ions by sodium cyanoborohydride at pH 1–2 before it could be further reduced to arsine by sodium borohydride at pH 1–2. Methylarsonic acid and dimethylarsinic acid were reduced to methylarsine and dimethylarsine respectively by the sodium borohydride reagent at pH 1–2.

The samples were placed in a reaction chamber and adjusted to the appropriate pH. The reagents were added and the volatile arsines produced were sparged from the sample with helium. The helium was passed through a U-tube, half packed with glass beads, that was cooled in liquid nitrogen, where the arsines were collected. When the reaction was complete the liquid nitrogen was removed and the U-tube was allowed to warm up to room temperature. The arsines volatilized from the trap in order of their boiling points. These are: arsine $-55\,°C$; methylarsine $2\,°C$; and dimethylarsine $55\,°C$. The helium was passed to an electric discharge where arsenic emission lines were produced. These emission lines were monitored at 234.9 nm or 228.8 nm by a monochromator-photometric read-out system, and this provided a very sensitive method of determining arsenic. Because the arsines volatilized at different times and hence passed through the detector at different times the read-out had the appearance of a chromatogram, the peaks corresponding to arsine, methylarsine and dimethylarsine. In this way the arsenic species in the sample were determined. The detection limit was 0.05 ng for As(III) and As(V) and 0.5 ng for the methylarsenic acids. Samples of natural waters, bird eggshells, seashells and human urine were analysed and the methylarsenic acids were found in all the samples. Typical results obtained are shown in Table 3.2.

TABLE 3.2
Species of arsenic found in natural water and human urine samples (from Braman and Foreback, Ref. 40)

Sample	As(III) ppb	As(V) ppb	Methylarsonic Acid ppb	Dimethylarsinic Acid ppb	Total ppb
Fresh water					
1	<0.02	0.25	<0.02	<0.02	0.25
2	<0.02	0.16	0.06	0.30	0.42
3	<0.02	0.27	0.11	0.20	0.68
4	<0.02	0.32	0.12	0.62	1.06
5	0.79	0.96	0.05	0.15	1.95
6	2.74	0.41	0.11	0.32	3.58
7	0.89	0.49	0.22	0.15	1.75
Saline water					
1	0.12	1.45	<0.02	0.20	1.77
2	0.62	1.29	0.08	0.29	2.28
3	0.06	0.35	0.07	1.00	1.48
Human urine					
1	<0.01	0.84	0.61	8.9	10.4
2	5.1	7.9	2.5	10.4	25.9
3	<0.5	2.4	2.4	25.2	30.0
4	2.4	4.3	1.8	15.5	24.0

Edmonds and Francesconi[41] point out that the determination of the methylated arsenicals can be carried out without the special analysis equipment used by Braman and Foreback, by using a 'conventional' hydride generation apparatus connected to an atomic absorption spectrophotometer via a U-tube cooled in liquid nitrogen. Another interesting point made by these authors is that in the past, to determine the total arsenic level in a sample, it was necessary to digest or ash the sample to convert all the arsenic to inorganic form. This was necessary because the determination consisted of generating arsine and measuring this arsine colorimetrically with silver diethyldithiocarbamate. This reagent does not react to the methylarsines, only arsine itself. However since in the atomic absorption method the arsine generated is converted to atomic arsenic for measurement, it reacts with equal sensitivity to arsine and the methylarsines. Hence the reduction of the sample with sodium borohydride (after reduction of As(v)) will result in all the arsenic forms being measured by the atomic absorption method, without the need for a prior digestion.

(Arsenite and arsenate in natural water samples have also been distinguished by their reaction with ammonium molybdate to form a molybdenum blue.[42] Arsenate reacts like phosphate; arsenite does not.)

METALS IN NATURAL WATERS

The metal pollutants that enter natural waters have an effect on the life in the water. One of the prime reasons for analysing samples and collecting data on pollutant levels must surely be to assess what effect certain activities are likely to have upon the waters that receive the pollutants arising from them. The effect that metals have is dependent on the particular form in which the metal exists in the water. Metals have an effect on, for example, the growth of algae. This effect is either as a stimulant, because the metal is a nutrient required by the plant, or as a depressant because the metal is toxic to the organism. In either case the metal must be physically available for the plant to absorb it through its surface. The most direct way in which the metal is available is when it is present as a metal ion in 'true' solution. If the metal is complexed with a complexing agent that the plant cannot metabolize, then it is not as readily available to affect the plant growth. Similarly if the metal is associated with the particulate matter in the water, either because the metal is in the form of a colloidal precipitate such as iron hydroxide, or if the metal ion is adsorbed on the particulate matter, then it does not have the same effect as the same amount of metal present as the free ion. These considerations form the basis on which the differentiation of the form of metals in natural waters is made, i.e. the amount available as a 'free' ion (labile metal), the amount that is complexed, and the amount that is associated with particulate matter of a certain size.

Labile metal determination

Metal ions in solution in a natural water are always complexed. This may be with simple inorganic ligands, such as water, carbonate, sulfate or halides, or with organic ligands such as amino acids, carboxylic acids, humic acids or tannins. Hence the determination of the 'free' ion is never exactly that, but is the determination of these complexes that are indistinguishable from the aqua complexes by the method of measurement. These are more commonly described as the 'labile' metal concentrations.

These determinations are commonly carried out using electrochemical techniques, since these are the only techniques that are available to make these measurements at the $\mu g \, l^{-1}$ levels at which many of the metals of interest occur. Using anodic stripping voltammetry (a.s.v.), the levels and species of cadmium, copper and lead in such waters as the Rouge River, Detroit River, Lake Erie and Lake Michigan were determined by Allen et al.[43] The technique has been described in Chapter 2 and the limitations as to the metals that can be determined are given there. However in the determination of particular species of metals the considerations are somewhat different to those when determining 'total' metal levels. These are discussed as follows by Chau and Lum-Shue-Chan[44] in a description of the method they used to determine the labile and strongly bound metals in lake water. The metals determined were Zn, Cd, Pb and Cu. It is important that the sample be analysed in a state as near to its original state as possible, so that some assurance is obtained that the results of the analysis are not artifacts introduced by the analytical procedure. However to carry out the analysis requires that some changes be made to the sample; these changes must be minimized in order to preserve the validity of the sample.

In the polarographic analysis it is necessary to use a supporting electrolyte to maintain a constant ionic strength in the solution during the plating–stripping processes. The pH of the solution must be controlled because the activity of the metal and the resulting oxidation current peaks are affected by changes in pH. Also the pH in the analysis should be as close as possible to the pH of the original sample since changes in pH could cause changes in the species present. A lowering of the pH would cause a shifting towards more metal ion liberation; raising the pH would reduce the metal ion activity because of precipitation, adsorption on colloidal material or an increase in the apparent strength of the complexes present. The oxygen in the solution must be removed since it interferes in the measurement. It is usually removed by bubbling nitrogen through the solution but this also removes some of the dissolved carbon dioxide from the sample and raises the pH. Hence it is necessary to add a buffer to the sample to maintain the pH in the same range as the original.

The supporting electrolytes that are used in 'regular' polarographic analysis are such materials as potassium chloride, ammonia–ammonium chloride, borax and tartrate. The complexing electrolytes such as tartrate change the concentration of the free metals, and the supporting electrolytes which are either acidic or basic are not acceptable, as discussed above. In addition zinc cannot be determined at low pH because of the interference from hydrogen gas evolution as the hydrogen ions are reduced. The anodic wave of copper does not appear in potassium nitrate solution and is only weakly defined in potassium chloride solutions. The supporting electrolyte which was found to be satisfactory by Sinko and Dolezal[45] for the analysis of Cd, Cu, Pb and Zn in natural waters was sodium acetate–acetic acid solution, which also acts as the buffer in the determination. This was the system used by Chau et al.[48] in their work on speciation. This buffers in the pH range 6.8–7.8, which is the range of the large majority of natural samples.

The acetate ion will form complexes with many metals. These complexes are weak, with stability constants of k_f value 1–3. With these low k_f values it is impossible to separate by polarographic measurement those metal ions which are aquo-complexed and those which are acetate complexed. Hence the labile metal determined in this case will include the metal in complexes which are acetate exchangeable.

In the sample solution there are complexes of the metals. These complexes are in equilibrium with the amount of metal ion produced by their dissociation. If the complexes are 'strong' this amount of metal is small, where they are weak the equilibrium concentration of metal ion is higher. The plating process involves the removal of metal ion from solution onto the electrode. The basis for the method is that the amount of metal removed from the solution is only a small fraction of the amount present, so that any equilibria existing in solution are not significantly disturbed by the removal. However it is necessary to remove a measurable amount of metal in order to carry out the stripping operation. If the metal ion is in excess over the equilibrium value of the complex, i.e. the stoichiometric ratio of metal to liquid is greater than one, then there is no disturbance of the equilibrium by the removal of some of the metal ions. If the stoichiometric ratio is one or less, then the removal of metal ions could change the equilibrium and result in more dissociation of the strong complexes. Hence in this technique it is important that the plating time be minimized, to avoid breaking down a significant fraction of the strongly bound complexes where the ligand concentration is the same as, or in excess of, the metal ion concentration.

To determine the total metal concentration in the sample the organic materials are destroyed by digestion and the 'labile' metal concentration is redetermined.

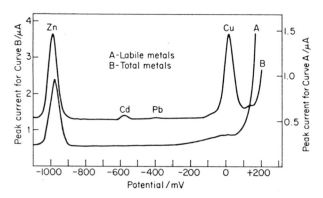

Fig. 3.3. Voltammograms showing the determinations of labile and strongly bound metals in a Hamilton Harbour sample. Sensitivity for labile metal, 2 μA full scale; for total metal 5 μA full scale. (From Ref. 44.)

In the absence of the organic complexing agents, this is the total metal concentration, and the difference between the two represents the strongly bound metals.

In the procedure used by Chau and Lum-Shue-Chan[44] the sample was filtered and 50 ml of the sample, buffered with sodium acetate, was plated. A hanging mercury drop was used and the deposition was carried out with stirring at -1.1 V for 3 min. The voltammogram was then obtained by scanning with a linear ramp from -1.1 V to 0.2 V at a rate of 5 mV s^{-1}. The total metal concentration was determined by the same procedure after the sample had been digested with a potassium persulfate–sulfuric acid digestion mixture.

The peak potentials of the four metals (relative to Ag/AgCl, saturated KCl electrode) in the system are: Zn, -0.965 V; Cd, -0.540 V; Pb, -0.362 V; Cu, 0.46 V. A typical voltammogram is shown in Fig. 3.3; some of the results obtained by Chau and Lum-Shue-Chan are shown in Table 3.3.

TABLE 3.3
Labile and strongly bound metals in some lakes in the Sudbury, Ontario, area (from Chau and Lum-Shue-Chan, Ref. 44)

Lake sample	pH	Zn (ppb)		Cd (ppb)		Pb (ppb)		Cu (ppb)	
		L[a]	SB	L	SB	L	SB	L	SB
Johnnie Lake	4.4	44	0	0.7	0	nd	6	6	24
Joe Lake	5.0	20	0	nd	nd	5	0	20	22
Kusk Lake	5.9	6	34	nd	0.3	nd	7	nd	24
Anderson Lake	6.2	4	1	nd	nd	nd	5	nd	6
Vermilion Lake West	6.5	4	2	nd	0.2	nd	7	nd	11
Wanapitae Lake	6.6	8	2	nd	nd	nd	4	nd	15
Lang Lake	6.8	3	8	nd	nd	nd	18	nd	12
Simon Lake	6.9	35	13	nd	nd	nd	3	nd	17

[a]L—labile, SB—strongly bound, nd—none detected.

There are some interferences with the procedure. Cyanide interferes at levels of 10 μg l^{-1} and higher, particularly in the determination of Pb and Cu, the peaks being distorted and shifted because of complex formation. Surfactants also interfere because they will adsorb onto the mercury drop and inhibit the electrode reactions so that the metal ions are not reduced to form the amalgam.

O'Shea and Mancy,[46] in studying simulated natural waters, carried out titrations of solutions of metal ions with organic ligands and solutions of ligands with metal ions. Anodic stripping voltammetry was used as the 'indicator' in these titrations and the measurement of both peak current and peak potential during the course of the titration enabled the differentiation of 'free' metal, labile metal complexes and non-labile metal complexes. The studies were made of the interaction of copper, thallium and cadmium with humic acid in a carbonate medium. The deoxygenation of the solutions necessary for the polarographic determination was carried out using a stripping gas consisting of a nitrogen and carbon dioxide mixture. This enabled the pH of the solution to be controlled.

At higher concentrations the labile metals may be determined directly using an ion selective electrode. Stiff,[47] using a cupric ion selective electrode, differentiated between the various species of copper in polluted natural water. However he reports that this system is only applicable where the total copper concentration is greater than 50 μg l^{-1} Cu, which would seem to rule out its use in most natural waters.

COMPLEXING CAPACITY

The complexing capacity of a natural water represents its potential for the complexing of metal ions by the materials in the water. If the stoichiometric ratio of a metal and a strong ligand is greater than one, then the metal exists in a labile state and also as a strongly bound complex. If the stoichiometric ratio is less than one, all the metal is strongly bound and there is an additional capacity in the water to convert metal ions to a non-labile state. This measurement is carried out by adding an excess of the metal ions of interest to the water and determining what fraction of the added metal is labile. It also measures the capacities for immobilizing metals other than by ligand formation such as adsorption on particulate matter.

Allen et al.[43] used anodic stripping techniques to follow the complexation of added copper to some natural waters. They determined the rate constant of the complexation, and believed that the knowledge of the rate constant would be useful in the characterization of the ligand types present in the waters.

Chau et al.[48] have used the same techniques to determine the complexing capacity of a number of lake waters for copper. Incremental additions of copper

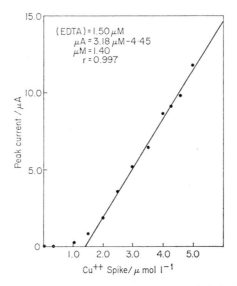

Fig. 3.4. Determination of the complexing capacity of EDTA (1.5 mol l⁻¹) in distilled water. r = coefficient of correlation. (From Ref. 48.)

were made and after an equilibrium period of a few hours the peak current due to the copper ion was recorded. A plot of the peak current versus the additions is made and the intercept that this makes on the 'addition' axis, when it is projected back to zero current, represents the complexing capacity of the water in terms of copper concentration. A plot of the results obtained when water containing EDTA is treated by the procedure is shown in Fig. 3.4. At the lower point of the graph it is seen that the line is curved and that some signal for copper was obtained even though there was an excess of EDTA present. This indicates that the determination of labile metal as described above does in fact determine some part of the metal that is complexed by strong ligand where the metal to ligand ratio is less than one.

In this work the copper binding capacity of several complexing agents of known stability constants was determined. Tartrate (log K $\sim$ 5), citrate (log K $\sim$ 6), glycine (log K $\sim$ 8), NTA (log K $\sim$ 13) and EDTA (log K $\sim$ 19) were examined. It was found that the techniques used measured the complexing by agents whose log stability constants were greater than about 13.

The measurement of the strong heavy metal chelating agents in water has been made by Kunkel and Manahan.[49] They treat the water with an excess of copper at a pH of 9.8–10.2. The uncomplexed copper is precipitated and after filtration the soluble copper is determined by atomic absorption spectroscopy. This represents the amount of strong chelating agents in the samples.

Ion exchange equilibrium techniques may also be used to determine the complexing capacity.[50] In this procedure incremental amounts of metal ion are

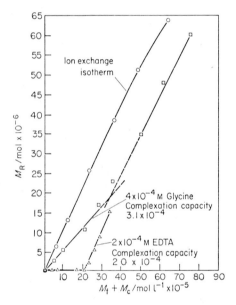

Fig. 3.5. Ion exchange equilibrium determination with two different complexing agents. (From Ref. 50.)

added to aliquots of the water containing the complexing agent. These samples are then equilibrated with an ion exchange resin. After equilibrium the concentration of the metal remaining in solution is determined and plotted against the amount of metal taken up by the resin. In the absence of a complexing agent there is a linear relationship between the amount of metal in solution and the amount taken up by the resin. In the presence of a complexing agent the amount of 'free' metal ions available to be taken up by the resin is reduced, so that a line of different slope is obtained. Hence, as the plots are made for the incremental additions of the metal ion, a line is obtained which changes slope at the point at which the complexing capacity of the sample is reached. The slope before this point is a function of the stability constant of the complexing agent. A plot for two complexing agents with different stability constants is shown in Fig. 3.5.

Size separations

The materials present in natural waters encompass a wide size range. There are the 'free' metal ions and the inorganic complexes which are perhaps a hundredth of a nanometer in diameter. Then there are organic complexes such as those of citric acid, oxalic acid, and amino acid similar to EDTA, which range up to a few tenths of a nanometer. The complexes with large organic molecules derived from degraded plant materials such as humic, fulvic and tannic acids, range up to a few nanometers in diameter. The metals such as iron and manganese can form hydrolytic polymers which range in size from 10 nm up to 100 nm. The hydroxyl groups on these hydrolytic polymers can form complexes with heavy metals and so bind them onto the polymer. There are also present finely divided clay particles, and other colloids on which metal ions can be adsorbed, and these may range in size up into the hundred nanometer range. Hence a separation of the materials in the water by size and a determination of the metal content of the various fractions can give some insight into where in the system the particular metals are located. In conjunction with the determination of the labile and total metal as described above it can give a good estimate of the metal species present. The limit on the usefulness of the technique is that there are gradations in size from one form to another and clear-cut differentiation is not possible. The techniques available for differentiation by size are filtration, ultrafiltration, dialysis and gel permeation chromatography.

Filtration using 'conventional' equipment will separate the metals such as suspended clays and the large hydrolytic polymers formed by metals such as manganese and iron. To separate materials smaller than this, ultrafiltration is used. In this the liquid is passed through a selectively permeable membrane with pressure. These membranes, made by the Amicon Corporation, are used in biological work to separate large molecules such as proteins, and their permeability is usually expressed on the basis of the molecular weight of the molecules that they will pass. They are available to pass materials of from less than 500 molecular weight up to 300 000 molecular weight. Smith[51] has reported on the use of the ultrafiltration technique to separate the organic materials in natural water by size, followed by determination of the complexing capacity of each fraction by a.s.v.

An alternative to ultrafiltration is dialysis. In this technique pure water is placed inside a dialysis bag and immersed in a sample of the natural water. When the volume inside the bag is small (e.g. 20 ml) compared to the volume of the sample (e.g. 250 ml) there is little shift in the equilibrium of the sample caused by this procedure. After equilibrium is reached, the water in the bag is removed for analysis of the metal concentration. This technique has been used by immersing the bags in the river or lake of interest and allowing the system to equilibrate for two weeks.[52]

A comparison of the results obtained on a model system by dialysis, an ion selective electrode, and a.s.v., has been made by Guy and Chakrabarti.[53] They examined the distribution of lead and copper in a humic acid solution as a function of pH by the three methods. Their results are shown in Fig. 3.6. It is seen that the three measurements give the same results. However, when a comparison was made of a.s.v. versus dialysis in determining the 'free' copper ion as the pH varied in a carbonate system, the dialysis method showed no change over the whole range pH 2–8, whereas a.s.v. showed a sharp drop at

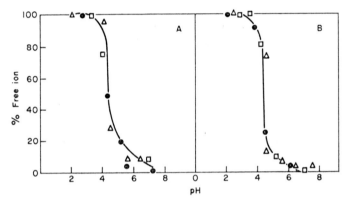

Fig. 3.6. A comparison of metal distribution in humic acid solution as a function of pH. A, 50 ng ml^{-1} copper + 10 g ml^{-1} humic acid. B, 50 ng ml^{-1} lead + 10 g ml^{-1} humic acid. ●—dialysis; △—ion selective electrode; □—a.s.v. (From Ref. 53.)

pH 6. Hence it is clear that when the species is of small size, such as $CuCO_3$ the dialysis technique does not make a separation, whereas a.s.v. can distinguish between Cu^{2+} (aq) and $CuCO_3$.

Gel permeation chromatography can also be used to separate materials on the basis of size. The gels used, such as Sephadex, are made with a closely controlled pore size into which the small molecules can diffuse but which the large molecules cannot enter. These are routinely used to separate organic materials on the basis of molecular weight. A scheme for speciating metals has been described by Burrell[54] which involves pH adjustment, metal chelation followed by solvent extraction, and gel permeation chromatography. One serious problem with the technique when applied to the classification of metal species is the adsorption of such materials as the Fe^{3+} ion and humic acid onto the gels.

The size discrimination aspect of metal species in natural systems is important. Florence and Batley,[55] in work on the removal of trace metals from sea water by a chelating resin found that the largest fraction of Cu, Zn, Pb and

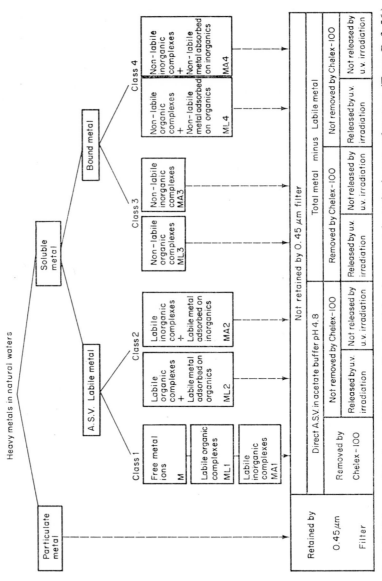

Fig. 3.7. Scheme of Batley and Florence for classifying heavy metal species in natural waters. (From Ref. 56.)

Cd was in a form that the resin would not extract. It was concluded that these metals were associated with colloidal particles of either inorganic or organic origin.

A scheme (see Fig. 3.7) for the classification of heavy metal species in natural waters has been described by Batley and Florence.[56] It was applied by them to the analysis of Cd, Pb and Cu in sea water. It uses a.s.v. as the means of determining the labile metal in the fractions examined. Differentiation is made between the labile metal and the metal bound in a strong complex by irradiating the sample with ultraviolet light. This breaks down the organic complexing agent and releases the metal. Measurement of the metals adsorbed by colloids is made by passing the sample through a Chelex resin column, where, as described above, the material on the colloids is not extracted, but the total soluble material is retained on the resin. Hence, by carrying out a.s.v. determination of labile and total metal on water samples that have either been subjected to u.v. irradiation, passed through a Chelex column, or both, and comparing the results with those obtained on an untreated sample, the classification shown in Fig. 3.7 may be determined.

The scheme divides the soluble metal into four classes: class 1 consists of such materials as citrate, glycinate, chloride and sulfate complexes; classes 2 and 4 comprise mainly metal ions adsorbed on colloidal species such as clays, silicates and organic detritus; class 3 consists of strong metal complexes of organic ligands such as humic acids.

REFERENCES

1. G. Westoo, *Acta Chem. Scand.* **20**, 2131 (1966).
2. S. Jensen and A. Jernelov, *Nature (London)*, **223**, 753 (1969).
3. J. C. Gage, *Analyst* **86**, 457 (1961).
4. H. O. Bouveng, *Determination of Methylmercury by Gas Chromatography*, Swedish Water and Air Pollution Laboratory Publication C7A, Stockholm, Sweden, 1968.
5. J. F. Uthe, J. Solomon and B. Grift, *J. Assoc. Off. Anal. Chem.* **55**, 583 (1972).
6. M. Fujiki, *Jpn Analyst* **19**, 1507 (1970).
7. R. Hartung, *The Determination of Mono- and Dimethylmercury Compounds by Gas Chromatography*, Conference on Environmental Mercury Contamination, Ann Arbor, Mich., 30 September–2 October, 1970.
8. W. H. Newsome, *J. Agric. Fd Chem.* **19**, 567 (1971).
9. J. O. G. Tatton and P. J. Wagstaffe, *J. Chromatogr.* **44**, 284 (1969).
10. G. Westoo, *Acta Chem. Scand.* **22**, 2277 (1968).
11. S. Nishi, Y. Horimoto and R. Kobayashi, *Identification and Determination of Trace Amounts of Organic Mercury*, International Symposium for Identification and Measurement of Environmental Pollutants, Ottawa, Ont., Canada, June, 1971.
12. J. Uthe and F. A. J. Armstrong, *Toxicol. Environ. Chem. Rev.* **2**, 45 (1974).
13. A. W. Andren and R. C. Harriss, *Nature (London)* **245**, 256 (1973).
14. J. J. Bisogni and A. W. Lawrence, *Environ. Sci. Technol.* **8**, 850 (1974).
15. K. Matsunaga and S. Takahashi, *Anal. Chim. Acta* **87**, 487 (1976).
16. Y. Umezaki and K. Iwamoto, *Jpn Analyst* **20**, 173 (1971).
17. Y. K. Chau and H. Saitoh, *Int. J. Environ. Anal. Chem.* **3**, 133 (1973).

18. Y. Kimura and V. L. Miller, *Anal. Chem.* **32**, 420 (1960).
19. A. A. Christie, A. J. Dunsdon and B. S. Marshall, *Analyst* **92**, 185 (1967).
20. D. G. Langley, *J. Water Pollut. Control Fed.* **45**, 44 (1973).
21. S. Jensen, *Nord. Hyg. Tidskr.* **50**, 85 (1969).
22. C. A. Bache and D. J. Lisk, *Anal. Chem.* **43**, 950 (1971).
23. J. E. Longbottom, *Anal. Chem.* **44**, 1111 (1972).
24. P. T. S. Wong, Y. K. Chau and P. L. Luxon, *Nature (London)* **253**, 263 (1975).
25. L. J. Snyder, *Anal. Chem.* **39**, 591 (1967).
26. W. W. Parker, G. Z. Smith and R. L. Hudson, *Anal. Chem.* **33**, 1170 (1961).
27. E. J. Bonelli and H. Hartmann, *Anal. Chem.* **35**, 1980 (1963).
28. V. Cantuti and G. P. Cartoni, *J. Chromatogr.* **32**, 641 (1968).
29. H. H. Hill and W. A. Aue, *J. Chromatogr.* **74**, 311 (1972).
30. W. A. Aue and H. H. Hill, *J. Chromatogr.* **74**, 319 (1972).
31. D. A. Segar, *Anal. Lett.* **7**, 84 (1974).
32. Y. K. Chau, P. T. S. Wong and H. Saitoh, *J. Chromatogr. Sci.* **14**, 162 (1976).
33. Y. K. Chau, P. T. S. Wong and P. D. Goulden, *Anal. Chim. Acta* **85**, 421 (1976).
34. C. S. Evans and C. M. Johnson, *J. Chromatogr.* **21**, 202 (1966).
35. J. Benes and V. Prochazkova, *J. Chromatogr.* **29**, 239 (1967).
36. L. Barkes and R. W. Fleming, *Bull. Environ. Contam. Toxicol.* **12**, 308 (1974).
37. V. Vlasakova, J. Benes and J. Parizek, *Radiochem. Radioanal. Lett.* **10**, 251 (1972).
38. Y. K. Chau, P. T. S. Wong and P. D. Goulden, *Anal. Chem.* **47**, 2279 (1975).
39. B. C. McBride and R. S. Wolfe, *Biochemistry* **10**, 4312 (1971).
40. R. S. Braman and C. C. Foreback, *Science* **182**, 1247 (1973).
41. J. S. Edmonds and K. A. Francesconi, *Anal. Chem.* **48**, 2020 (1976).
42. D. L. Johnson and M. E. Q. Pilson, *Anal. Chim. Acta* **58**, 289 (1972).
43. H. E. Allen, W. R. Matson and K. H. Mancy, *J. Water Pollut. Control Fed.* **42**, 573 (1970).
44. Y. K. Chau and K. Lum-Shue-Chan, *Water Res.* **8**, 383 (1973).
45. I. Sinko and J. Dolezal, *J. Electroanal. Chem.* **25**, 299 (1970).
46. T. A. O'Shea and K. H. Mancy, *Anal. Chem.* **48**, 1603 (1976).
47. M. J. Stiff, *Water Res.* **5**, 585 (1971).
48. Y. K. Chau, R. Gachter and K. Lum-Shue-Chan, *J. Fish. Res. Board Can.* **31**, 1515 (1974).
49. R. Kunkel and S. E. Manahan, *Anal. Chem.* **45**, 1465 (1973).
50. M. L. Crosser and H. E. Allen, *Soil Sci.* **123**, 176 (1977).
51. R. G. Smith, *Anal. Chem.* **48**, 74 (1976).
52. P. Benes and E. Steinnes, *Water Res.* **8**, 947 (1974).
53. R. D. Guy and C. L. Chakrabarti, *Chem. Can.* **26** (November 1976).
54. D. C. Burrell, *Atomic Spectrometric Analysis of Heavy Metal Pollutants in Water*, Ann Arbor Science, Ann Arbor, Mich., 1974, p. 96.
55. T. M. Florence and G. E. Batley, *Talanta* **23**, 179 (1976).
56. G. E. Batley and T. M. Florence, *Anal. Lett.* **9**, 379 (1976).

CHAPTER 4

ANALYSIS OF
INORGANIC MATERIALS

INTRODUCTION

This chapter is concerned with the analysis of the non-metallic inorganic pollutants in environmental samples that are brought to the laboratory. For air analysis the methods described are those that are used on the filters or on the solutions obtained by absorbing the particular material from the air in an absorber or impinger. (The direct-reading methods for the air pollutants are described in Chapter 8.) In the case of sediments and soil, the solutions are obtained by extracting the samples with a variety of aqueous extractants.

Most of the methods described involve carrying out some chemistry on the sample solution and then measuring the concentration of the resulting material either colorimetrically or with an ion-selective electrode. In the majority of laboratories carrying out environmental analysis, these manipulations are carried out in automated equipment. This is because of the large number of samples that can be handled, the lower manpower requirements, and also because of the greater precision obtainable.

The laboratory automation techniques have their greatest use in the clinical laboratory. At a hospital the large number of similar samples each requiring complex analyses makes for a perfect field for automation. The automated systems used in environmental analysis are for the most part adaptations of those developed in the clinical field.

The automated processes can be divided into two types: the discrete sample analyser and the continuous-flow system.

Discrete sample analysers

The discrete sample analyser acts like a laboratory robot in duplicating manual procedures. The sample is metered into a container of some sort (either a tube or a dish) and operations are carried out on this sample as it is carried through the system.

The systems are commonly capable of handling several hundred samples per hour. The most complex part of these systems is the mechanical operation of the syringes, used for reagent additions and sample transfer, because of the stringent demands for high-precision and high-speed operation. The operations are carried out in open tubes and hence the processing is usually restricted to reagent addition, heating, cooling and filtering. Therefore the determinations possible are those in which the final measurement, most commonly a colorimetric one, can be made in the sample matrix. This restriction probably accounts for the fact that, despite their high sample-handling capability, discrete sample analysers do not appear to be widely used in environmental analysis. In many analyses it is desirable to separate the parameter being measured from the sample matrix by, for example, distillation, to overcome interferences and to provide a concentration step for the final measurement. This is not possible in the open vessel systems. However, the development of more selective and sensitive final measuring processes, such as the fluoride ion selective electrode, may enlarge the area of use of this type of automated analyser.

Continuous-flow analysers

Continuous-flow analysis was introduced commercially in 1957 in the clinical field by the Technicon Corporation with the designation Auto Analyzer®. These analysers came into wide use in clinical laboratories and in analytical laboratories generally. The majority of automated methods published for analysis of environmental samples use Auto Analyzer® equipment. The key to the

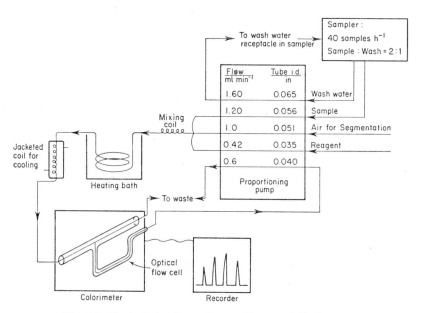

Fig. 4.1. Typical simple continuous-flow analytical system.

continuous-flow system is that the liquid flowing through the tubes is segmented by a series of air bubbles. The segment of liquid between two bubbles is prevented from mixing to any great extent with the liquid the other side of the bubbles so that each segment can be carried through the system as an integral portion. A typical system is shown in Fig. 4.1.

The liquids are pumped with a peristaltic pump. This consists of a series of rollers which are drawn over a platen by two chains. A plastic tube is squeezed between the rollers and the platen, and the fluid in the tube is pumped by the forward action of the rollers. The flow rate is changed by using pump-tubes of different internal diameters; flow rates from 0.15 ml min^{-1} to 3.9 ml min^{-1} are obtainable in 20 steps.

The sampler consists of a circular tray that holds 40 sample tubes. Samples are taken into the system via a sample arm that aspirates the sample for a certain time, then aspirates a wash solution before the next sample. The sampling rate varies according to the analysis being carried out, typically it is 20–60 samples per hour. The sample time:wash time ratio is also varied; typically it is from 2:1 to 1:2. For samples containing some suspended solids, an agitator may be used in the sample tube as the sample is aspirated.

Reagents are added to the sample by being pumped through the appropriate size pump tube. Mixing is effected by passing the liquid through a coil, placed with its axis horizontal. As the segments of liquid pass around the turns of the coil the heavier phase moves through the segment, to the back of the segment on the upward side of the coil, to the front on the downward side and to each side of the segment on the top and bottom parts of each turn of the coil. Temperature control is obtained by passing the liquids through coils in a bath at the desired temperature. To carry out distillations the liquid is passed through a coil in a heating bath at a temperature perhaps 30 °C above the boiling point of the liquid. The boiling liquid coming from this coil flows downwards over a ribbed vertical surface where the vapour produced separates.

Such operations as solvent extraction are carried out by passing a mixture of the solvent and the aqueous sample through a coil filled with small glass beads to carry out the mixing of the phases and the extraction. Separation of the two phases is carried out by gravity, aided by 'directing strips' of an appropriate material. To help separate a solvent heavier than water, such as chloroform, a strip of Teflon® is laid on the bottom of the separating tube. The chloroform will wet Teflon® but water will not, so that, as the bubbles of chloroform settle to the bottom of the tube, they coalesce on the Teflon® strip and run in a continuous stream along the tube.

Filtering is carried out by dropping the sample to be filtered onto the top of a continuously moving tape of filter medium. The liquid passes through the filter medium into a chamber from which it is pumped back into the system. By installing a mixing chamber directly above the filter medium, precipitation reactions can be carried out without the problem of flushing the solids produced through tubes. If the filter medium is made of a hydrophobic medium the filter system can be used to separate solvents and water, since the solvents will wet the medium and pass through, while the water will roll in drops off the top of the medium.

To separate soluble material from colloids, dialysis can be used. The sample flows on one side of a membrane and a clean reagent solution flows on the other side to collect the ions that pass through. By using a gas permeable membrane a separation can be made of gases from the liquid sample. Such a system is used, for example, to remove CO_2 from a sample into another stream where the pH change from the absorbed CO_2 can be measured.

In many of the applications of the Auto Analyzer® the final measurement is a colorimetric one. However, the Auto Analyzer® has been interfaced with almost every measuring device that exists. It is commonly used to treat and present samples for atomic absorption spectroscopy. When the final measurement is colorimetric the sample is 'de-bubbled' and the liquid drawn through an absorption cell.

The early Auto Analyzers®, now designated 'Auto Analyzer I'®, used mixing coils and connecting tubing of internal diameter up to about 2.4 mm. The de-bubbling in the colorimeter allowed successive segments of liquid to coalesce and the air bubble to separate. The flowcell in the colorimeter was of comparatively large diameter and the sample rate was limited by the loss of integrity of the sample, i.e. at high sampling rates the peaks obtained would overlap each other. In the new versions of the Auto Analyzer® (designated 'Auto Analyzer II'®) the diameter of the tubing is reduced, and the air pumped to segment the liquid represents a much higher fraction of the total flow (i.e. 40% vs 10% for AA I). The separation of the liquid to pass through the flow cell is made by drawing the liquid from the bottom of a T-joint where the surface tension of the liquid keeps a continuous flow down the vertical arm while the air bubbles pass through along the horizontal arm (see Fig. 4.1). Hence successive segments of liquid are very small and they do not coalesce in the de-bubbling operation. The flowcell diameter is also small and the net result is that the integrity of the sample is maintained much better in the AA II system than in the AA I, resulting in higher sample rates. These characteristics are discussed by Snyder et al.[1]

Because all the samples and standards are treated in exactly the same way, the precision of the results obtained is much better than that obtained by manual operations. Another factor, which is particularly important when measuring at 'trace' levels, is that the reagents are being continuously added so that separate blank determinations are not necessary. Any contamination brought in with the reagents is added continuously, both to the samples being measured and to the standards used to calibrate the system, and hence it disappears into the baseline. The part that is critical in the measurement of trace levels is that the standards themselves are of the nominal concentration. In the author's experience, carrying out an analytical measurement with an automated system results in a detection limit of about one-tenth of the detection limit obtainable in a manual determination using the same chemistry.

ANALYTICAL METHODS

Boron

Boron may be determined by atomic spectroscopy but for this element it is a very insensitive technique. Boron is commonly measured in water samples by a variety of colorimetric or fluorimetric methods. Carmine is used when boron levels are in the 1–10 mg l^{-1} B range.[2] The sample is evaporated under alkaline conditions and any organic matter is destroyed by ashing. The residue is dissolved in a small volume of hydrochloric acid and reacted with carmine in sulfuric acid. A bluish-red colour is developed whose absorbance is measured at 585 nm. The carmine method has been automated.[3] The detection limit of this automated method is 0.02 mg l^{-1} B.

For boron levels in the 0.1–1.0 mg l^{-1} B range, the curcumin method is used.[2] In this the sample is mixed with curcumin reagent and evaporated to dryness in a small evaporating dish which is floated on the surface of a water bath at 55 °C. The time of this drying is maintained constant (80 min is recommended) for both standards and samples. After this time the dishes are removed and the rosocyanine dye which is produced is dissolved in ethyl or isopropyl alcohol and made up to volume (25 ml). The absorbance at 540 nm is measured. The method suffers interference from precipitation of hardness salts in the alcohol solution. Hence for hard water the sample is passed through a strongly acidic cation exchange resin to remove the calcium and magnesium. Nitrate above 20 mg l^{-1} N will also interfere in the method. To facilitate analysis of waters containing high levels of nitrate this has been removed by reducing the nitrate to ammonia using sodium hydroxide and a suspension of finely ground aluminium powder.[4] In a modified curcumin method,[5] the sample is acidified and the boron extracted into 2-ethyl-1,3-hexanediol in chloroform. The chloroform extract is then treated with curcumin in acetic acid, and a small amount of sulfuric acid. The rosocyanine dye forms and is dissolved in ethyl alcohol. With this procedure the interferences from nitrate and hardness salts are eliminated as is any interference from fluoride. There is also no need to carry out the evaporation step in order to form the rosocyanine.

The curcumin reaction has also been used in an automated method for boron in sea water using a discrete sample analyser, the AutoLab®.[6] In this procedure the sea water sample is taken and the water removed by reaction with propionic anhydride. This reaction is catalysed by oxalyl chloride. The curcumin is then added as a solution in acetone, together with an acetic acid–sulfuric acid mixture. An ethanolic solution of ammonium acetate buffer is then added and the absorbance measured.

Another colorimetric reaction for the determination of boron is that with 1,1′-dianthrimide.[7] In this procedure the sample (5 ml) is mixed with sulfuric acid (1 ml) and evaporated to 1 ml or less in an oven at 90 °C. Overnight evaporation is usually sufficient to dehydrate the solution. To the dehydrated sample is added a solution of dianthrimide in sulfuric acid (5 ml) and the mixture is incubated at 90 °C for a further three hours. The solution is cooled and the blue colour developed is measured at 620 nm. Fluoride and nitrate do not interfere in this procedure since they are volatilized from the solution during the dehydration. Organic materials do interfere in that they produce colours as they char in the sulfuric acid. Organic materials are removed by treating the sample with hydrogen peroxide before the dehydration step. An alternative procedure that has been found more convenient in handling samples containing high levels of organics is to add solid potassium persulfate to the dehydrated sample before the dianthrimide addition.[4] In this work an automated method for measuring the colour in the 'manually dehydrated' sample is described. The detection limit of the manual method is about 20 mg l^{-1} B. Of the automated method it is 10 mg l^{-1} B.

A modification of the dianthrimide method is described by Levinson.[8] In this procedure the sample is evaporated to dryness. Calcium hydroxide is added to the sample before the evaporation and this prevents loss of boron. The dried sample can then be reacted with the dianthrimide solution in sulfuric acid. In this way the overnight dehydration step is avoided and a larger sample volume can be taken and rapidly evaporated.

A method using solvent extraction of the ferroin ion-association salt of borodisalicylate is described by Bassett and Matthews.[9] The sample is adjusted to pH 5.5 and sodium salicylate and sulfuric acid added. Borodisalicylate is

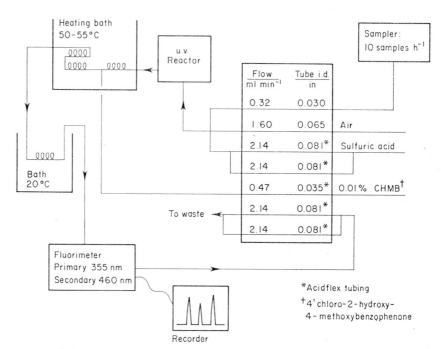

Fig. 4.2. Manifold for determination of boron by fluorescence.

formed and this is reacted with ferroin. The ion-association salt is extracted into chloroform, the extract is washed with water and the absorbance at 516 nm is measured. The reaction to produce the borodisalicylate is slow; a one hour period is used for this. Interferences from copper and zinc are removed by the addition of EDTA and diammonium hydrogen phosphate respectively. The method measures boron over the range 0.01–2.7 mg l^{-1} B.

A method on the Auto Analyzer® has been used by James and King.[10] This uses the decrease in absorbance of a solution of disodium chromotropic acid by boric acid. The absorbance at 362 nm is monitored in a solution buffered at pH 7 with sodium acetate. The detection limit is 0.1 mg l^{-1} B.

An automated method using the fluorescence of the species formed by 4'-chloro-2-hydroxy-4-methoxybenzophenone (CHMB) and boron has been described.[11] The reaction is carried out at 50–55 °C in a 90% sulfuric medium; the excitation is at 355 nm. The fluorescence emission is measured at 460 nm. The detection limit = 1 mg l^{-1} B. The only significant interferences are organic materials and these are destroyed by ultraviolet irradiation of the sample prior to formation of the fluorescent species. The manifold used in this method is shown in Fig. 4.2.

Water samples

It is believed that preservatives are not needed when sampling water for boron analysis. It should be remembered, however, that since nitrate is a potential interference in many of the methods, samples which have been preserved by the addition of nitric acid, for example, those samples taken for metal analysis, should not be used for the determination of boron.

Sediment–soil samples

The boron in soil or sediment samples is often classified as the water-soluble, acid-soluble, and total, boron. For water-soluble boron the sample is boiled with water or continuously extracted in a soxhlet thimble. The acid-soluble boron is determined by extraction with 85% H_3PO_4. For a total determination an alkaline fusion of the solid is made.

Biological samples

Biological samples are ashed and the residue dissolved in the acid appropriate for one of the determinations above. The ashing must be carried out under alkaline conditions to avoid loss of boric acid. Most vegetation is sufficiently alkaline to prevent loss of boron on ignition. Calcium hydroxide may be used to ensure alkalinity.

Chloride

Chloride in solution may be measured by titration, colorimetrically or with a chloride ion-selective electrode. The chloride may be titrated with silver nitrate or with mercuric nitrate. When titrated with silver nitrate, potassium chromate is used as an indicator. When chloride is in excess, silver chloride is precipitated; when silver is in excess, red silver chromate is formed. In the titration with mercuric nitrate, the soluble slightly dissociated mercuric chloride is formed. The indicator used is diphenylcarbazone which with mercuric ion forms a purple complex. In this determination pH control is critical. The chloride can also be titrated with silver nitrate using a silver–silver chloride electrode to determine the end-point. These procedures are described in *Standard Methods*[2] and the ASTM standards.

The common colorimetric method uses the reaction with mercuric thiocyanate to liberate thiocyanate. This thiocyanate is then determined with ferric iron as the red ferric thiocyanate. The absorbance of this is measured at 480 nm.[12] The manifold used for the determination on the Auto Analyzer® is shown in Fig. 4.3. The detection limit of the method is 200 µg l[−1].

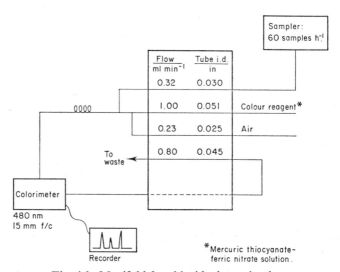

Fig. 4.3. Manifold for chloride determination.

Chloride may also be measured by reaction with mercury chloranilate to liberate chloranilic acid. The absorbance is measured at 332 nm in a similar way to that used in analysis of sulfate.[13]

Another automated colorimetric method which may be used is that employing chromotropic acid.[14] In this method chloride is used to catalyse the reduction of nitrate to nitrite. The nitrite reacts to form a coloured species with chromotropic acid which is measured at 505 nm. The reaction is carried out in 54% sulfuric acid. The detection limit of the method is 0.25 mg l[−1].

Chloride may also be determined with a chloride ion-selective electrode.[15] With the commercially available electrodes[15a] which use silver halide–silver sulfide membranes, levels of chloride down to 1 mg l[−1] may be measured. An improved chloride electrode has been described by Sekerka et al.[15a,16] which uses a membrane of mercuric sulphide–mercurous chloride. This has a higher sensitivity than the electrodes based on silver salts and can be used to measure concentrations down to 20 µg l^{-1} Cl.

Interference in most of these methods comes from the other halides, i.e. bromide and iodide, which are measured as chloride, and from the silver- and mercury-precipitating anions. Usually, in most environmental samples none of these interferences are significant. In the case of the measurement by the ion-selective electrode, sulfide is potentially a severe interference since it 'poisons' the membrane of the electrode. For this measurement sulfide can be oxidized with hydrogen peroxide or precipitated as bismuth sulfide.

Water samples may be analysed as they are. No chemical preservatives are added to the sample. Sediment and soil samples are extracted. Vegetation is frequently analysed for chloride in a similar manner to that used for fluoride. The ashing-extraction procedure is similar to that used in fluoride analysis; a recent paper[17] recommends an oxygen-flask combustion in a simultaneous determination of sulfate, chloride and fluoride. Chlorides in air samples may be collected in distilled water and this solution analysed by one of the above methods.

Bromide

Bromide may be determined by oxidizing it to hypobromite and then reacting this with phenol red. The phenol red forms an indicator similar to bromophenol blue and the concentration of the original bromide is determined by measuring the absorption of this indicator at 590 nm. The oxidation is carried out using chloramine T (sodium toluene-p-sulfonchloramide). The concentration of the chloramine T and the time of reaction are critical factors. The reaction is stopped after a certain time, usually 20 min, by the addition of sodium thiosulfate which destroys the chloramine T. A buffer is added at the beginning to control the pH at which the colour is read to 5.0–5.4. The absorbance is measured at 590 nm. A calibration curve is prepared against which the samples are determined. The minimum detectable concentration is 100 µg l^{-1} Br. This procedure is described in *Standard Methods*.[2]

Another colorimetric determination for bromide is based on the catalytic effect of bromide ion on the rate of oxidation of iodine to iodate by potassium permanganate in sulfuric acid solution.[18] The reaction is allowed to proceed for a certain time under controlled conditions of pH, temperature and concentration of the reacting materials. The reaction is then stopped by extracting the unreacted iodine with carbon tetrachloride. The amount of unreacted iodine is determined by measuring the absorbance at 515 nm. This amount is inversely proportional to the concentration of the bromide ion present. The interferences in this method are those materials which oxidize iodine or reduce iodate or permanganate under the conditions of the test. Free chlorine constitutes the worst interference and must be removed by boiling or by purging with air.

Nitrite at a concentration greater than $1 \mu g \, l^{-1}$ is a negative interference because it will reduce iodate. Some metal ions cause interference but most natural waters do not contain these materials in high enough concentrations to cause problems. The reaction is normally carried out in an ice bath to obtain a controlled temperature. The detection limit of the method is $1 \, mg \, l^{-1}$ Br.

A titrimetric method may be used for bromide where the concentrations are high, i.e. $50 \mu g \, l^{-1}$ or above. The method also determines iodide. In this method the iodide in the sample is oxidized to iodate with bromine. This reaction is carried out in a buffered solution. After it is complete, the excess bromine is destroyed by reaction with sodium formate. The iodate is then reacted with potassium iodide to liberate iodine. This iodine is determined by titration with sodium thiosulfate. The combined iodide and bromide in the sample are then determined by oxidizing them to iodate and bromate respectively using hypochlorite. The excess hypochlorite is destroyed with sodium formate solution and the combined bromate–iodate used to liberate iodine from potassium iodide. This iodine is determined by titration with sodium thiosulfate. The bromide concentration is calculated by subtracting the value for the iodide concentration found above from the combined bromide–iodide value. Bromide may also be determined using a bromide ion selective electrode.[15a,19] The discrimination against chloride ion is good; it will tolerate a Cl:Br ratio of up to 400:1. This may not be satisfactory for natural water samples but is very useful in analysing air samples where bromide levels are often high from the lead scavengers used in gasoline.

Sample collection is the same as for chloride.

Iodide

Iodide is measured using the fact that iodide ions catalyse the reduction of ceric ions by arsenious acid. The direct determination of the loss of the colour of the ceric ion is difficult without a special recording device because the colour fades rapidly. However, by the addition of ferrous ammonium sulfate the reaction is stopped at a specific point in time and the remaining ceric ions then oxidize ferrous ions to ferric. The amount of ferric ion produced, which is proportional to the amount of ceric ion that remains, is then determined colorimetrically as ferric thiocyanate at 510 nm. A constant-temperature water bath at 30 °C is used to control the temperature of the ceric-reducing reaction. Chloride ion affects the reaction by acting as a sensitizer. To overcome the effect of varying chloride levels in the samples, an excess of sodium chloride is added to ensure that the reaction is carried out on the plateau of maximum chloride sensitization. The formation of non-catalytic forms of iodine and the inhibiting effects of silver and mercury are reduced by this addition.

This test measures only the iodide ions in solution. If other forms of iodine are present and it is desired to measure these too, then a digestion procedure must be carried out. A digestion with chromic acid and separation of the iodide by distillation is commonly used. The detection limit is about 0.5 µg l^{-1}. The procedure is described in *Standard Methods*.[2]

Sample collection is as for chloride. Iodide can also be determined with an ion-selective electrode.[15a]

Fluoride

Fluoride ion in solution may be measured titrimetrically, colorimetrically or by an ion-selective electrode. Before these procedures can be carried out, it is usually necessary to overcome the possible interferences from other ions in the sample. This may be done by separating the fluoride by distillation or by the addition of masking agents.

In the titrimetric methods the fluoride is titrated with thorium nitrate using as indicator alizarin red S, chrome azurol S or SPADNS trisodium salt of 4,5-dihydroxy-3-(p-sulfophenylazo)-2,7-naphthalene disulfonic acid). The colorimetric procedures may be direct or indirect. In the direct method a blue complex is formed between alazarin complexon, fluoride and a rare earth cation—either lanthanum[20] or cerium.[21] The concentration of fluoride is determined directly by measuring the absorbance of the lanthanum fluorochelate at 615 nm. This procedure will measure levels down to about 20 µg l^{-1} F. An automated procedure using the Auto Analyzer® is available in which the fluoride is distilled from the sample and the absorbance of the lanthanum fluorochelate measured at 620 nm.[22] The manifold is shown in Fig. 4.4. This has a detection limit of 40 µg l^{-1} F. For a more sensitive determination the chelate can be extracted into isobutyl alcohol containing hydroxylamine hydrochloride, the absorbance being measured at 570 nm. In the indirect method the fluoride is added to a solution of zirconyl SPADNS dye and competes with the SPADNS for the zirconium. The decrease in absorbance of the zirconyl SPADNS solution at 570 nm is then a measure of the fluoride concentration.

The fluoride ion-selective electrode is perhaps the most sensitive and selective of all the ion-selective electrodes.[15a] In a large number of laboratories it has practically replaced the colorimetric and titrimetric methods for fluoride analysis. It suffers interferences from hydroxyl ions, polyvalent ions that complex fluoride (aluminium, silicon and iron) and the effect on activity of the total ionic concentration. These interferences are overcome by the addition of a buffer that contains a sequestering agent for the polyvalent cations and also acts as a total ionic strength adjustment. With this electrode fluoride levels down to 5 µg l^{-1} F in solution may be measured manually.

The early total ionic strength adjustment buffer (TISAB) used citrate as the sequestering agent; the later modification using 1,2-diamino-cyclohexane-N,N,N',N'-tetraacetic acid (DCTA) in place of the citrate gives improved sensitivity.[23] An automated system using a Fisher Titralyzer® turntable has been described by Sekerka and Lechner.[24] This method has a detection limit of 2 µg l^{-1} F. In this work it was found that some of the reagents used to make the TISAB contained fluoride and this was one problem in obtaining a low

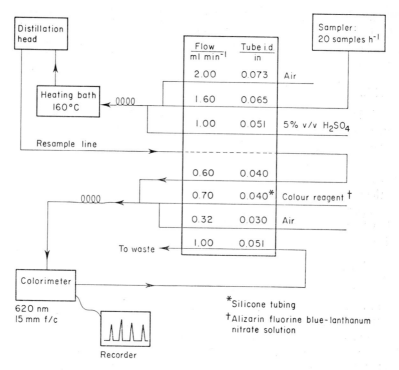

Fig. 4.4. Manifold for fluoride determination.

detection limit. Sodium chloride is normally used in preparing TISAB and this reagent in particular is always contaminated with some fluoride. It was found that an essentially fluoride-free TISAB was made using only sodium hydroxide and acetic acid. This was prepared in a concentrated form to minimize the dilution of the sample. These modifications, and the reproducibility of the procedure using automation, resulted in the low detection limit. The fluoride electrode can also be used in Auto Analyzer® equipment.

The fluoride electrode does not respond to fluoroborate ion (BF_4). Fluoroborates are used in the electroplating industry and if it is believed that a sample may contain fluoroborates from such a source it is necessary to distill the sample from acid to liberate the fluoride. The concentration in the distillate may then be measured by the electrode.

Water samples

No chemical preservatives are added to water samples to be analysed for fluoride. Polyethylene sample bottles are usually preferred to glass bottles. For the measurement the sample may be used directly if the fluoride electrode is used (but see reference above to fluoroborates). In the colorimetric and titrimetric measurements the fluoride is generally distilled from sulfuric acid.

Air samples

In air sampling the gaseous fluoride is normally caught in sodium hydroxide using an absorber or impinger. Fluorides in air particulates are brought into solution with sodium hydroxide and any organic material is destroyed by ashing, under alkaline conditions to prevent loss of hydrogen fluoride.

Biological samples

Samples such as vegetation are ashed under alkaline conditions to prevent loss of fluoride, or the ashing is carried out in an oxygen flask. The fluoride is separated by distillation of the ashed material with perchloric acid or the ashed material is solubilized and the fluoride measured directly with an electrode. (Perchloric acid is used for the distillation of samples containing low levels of fluoride because sulfuric acid itself contains trace levels of fluoride.)

Plants are more sensitive to fluoride than to most of the other air pollutants. Animals which graze on fluoride-contaminated forage can suffer injury. Hence vegetation is frequently sampled as a means of diagnosing such injury and also as a measure of air quality. There has been considerable discussion in the literature as to the best way to determine fluoride in vegetation, the techniques are reviewed by Cooke et al.[25]

Cyanide

Cyanide ions in solution can be measured titrimetrically, with an ion-selective electrode,[15a] or colorimetrically.[2] The titrimetric method is used where levels of cyanide greater than 1 mg l^{-1} CN are present. In this method the cyanide solution is titrated with silver nitrate solution. When cyanide is in excess the soluble $Ag(CN^-)_2$ complex is formed. When there is a small excess of silver this is detected by the indicator, paradimethylaminobenzalrhodanine, which with silver ion turns from yellow to pink.

The cyanide ion-selective electrode is usually used to determine cyanide by the 'known-addition' technique.[15a] A calibration curve is first obtained using standard solutions of cyanide. The shape of this calibration indicates whether the electrodes and millivoltmeter are operating properly. The potential obtained with the sample solution is then measured and small volumes of standard cyanide solutions are added incrementally to the sample. The potential obtained after each addition is recorded; the slope and intercepts of the plot of potential vs addition, enable the original concentration to be determined with greater accuracy than by a single potentiometric measurement. This technique will measure levels down to 2 µg l^{-1} CN.

In the colorimetric method used for cyanide, the cyanide is converted to cyanogen chloride by reaction with chloramine T. The cyanogen chloride forms a red-blue dye when reacted with a pyridine–barbituric acid reagent. The concentration of cyanide is determined by measuring the absorption at 578 nm. The detection limit of this method is 1 µg l^{-1} CN.

These methods suffer interferences from materials often present in water samples such as sulfide, halides, organic materials, etc., and it is usually necessary to separate the cyanide from the sample matrix. This is normally carried out by distilling the cyanide as hydrogen cyanide. (In the case of air samples collected by absorbers or impingers in sodium hydroxide solution, the distillation step may be unnecessary.)

This separation step introduces considerable complication into the determination of cyanides. The concern for cyanide pollution is that it is extremely toxic to many forms of aquatic life. This toxicity is due to the molecular hydrogen cyanide formed by the dissociation of the metal cyanides present in water. The cyanides exist in water mostly in two forms: as simple cyanide such as sodium cyanide, NaCN; or complex cyanides, such as those of zinc, $Na_2(Zn(CN)_4)$; iron, $Na_4(Fe(CN)_6)$; and cobalt, $Na_3(Co(CN)_6)$. The dissociation constants of these various metal complexes are different; hence the amount of free cyanide ion, and molecular hydrogen cyanide, produced is different. The alkali metal cyanides and the complexes of zinc, cadmium and lead dissociate readily. Lower dissociation is found with complexes such as nickel and cobalt and no measurable dissociation is found with the iron complexes. The desired analytical procedure is to be able to determine the free cyanide ion in the sample, the problem in the distillation step is to transfer this free cyanide ion to the distillate.

Historically the procedure has been to determine the 'simple' cyanide and the 'total' cyanide. The simple cyanide was obtained by distilling the sample with an organic acid. The total cyanide was obtained by refluxing the sample for an hour with sulfuric acid and a catalyst. In the distillation with an organic acid it is not certain how many of the 'weak' complex cyanides would liberate hydrogen cyanide. In the total cyanide refluxing, the iron cyanides are decomposed and liberate their hydrogen cyanide. However, the cobalt cyanides are not recovered completely, because of the catalytic decomposition of cyanide in the presence of cobalt in a strong acid solution at high temperature. A further complication is that thiocyanate decomposes to some degree in the process to yield cyanide.

The procedure which is now recommended, e.g. in the 14th edition of *Standard Methods*,[2] is to determine the 'total' cyanide on a portion of the sample by refluxing with sulfuric acid and a catalyst for an hour and collecting the HCN evolved. (In this procedure the cobalt cyanides are still not completely recovered.) A portion of the sample is then chlorinated, this destroys the cyanides which are present as dissociable cyanide ion. The cyanide remaining is then determined by a repeat of the one-hour refluxing operations. The cyanide in the distillate may be determined colorimetrically or with the electrode. The difference between the two results is expressed as the 'cyanide amenable to chlorination'. In the absence of interferences the determination can be made by carrying out the chloramine T, pyridine–barbiturate acid colour forming step on the original sample.

The interferences which are of concern are oxidizing agents, sulfide and thiocyanate. Oxidizing agents, which would destroy the cyanide in the storage and manipulation of the sample, are tested for when the sample is taken. If found they are removed by the addition of ascorbic acid or other reducing agent. Sulfide, which will distill as H_2S, interferes in the potentiometric, titrimetric and colorimetric determination. It is removed by precipitation as cadmium or bismuth sulfide or as sulfur with sodium sulfite. The sulfide should be removed when the sample is taken since the preservation technique used for cyanide is to bring the sample to a pH above 12. At this pH sulfide will react with cyanide to form thiocyanate. Thiocyanate forms cyanogen chloride with chloramine T and hence interferes in the colour reaction. It does not distill, but in the refluxing process it is partially converted to cyanide if the 'regular' cuprous chloride catalyst is used. Hence if thiocyanate is believed to be present, magnesium chloride is used as the catalyst in place of cuprous chloride. Thiocyanate may be determined by reaction with ferric iron under acid conditions to form ferric thiocyanate whose absorption is measured at 480 nm.

An alternative approach for the determination of total cyanides is the use of ultraviolet irradiation to break down the complex cyanides. The hour-long refluxing, to break down the complex cyanides, as recommended in *Standard Methods*, is inconvenient to duplicate in continuous-flow automated equipment. There is a method reported[26] in which a digestion in a heating bath is used. It has been found by the author and others[27] that irradiation with ultraviolet light rapidly breaks down complex cyanides and enables complete recovery to be made even from the iron and cobalt complexes. The ultraviolet irradiation is easy to carry out with Auto Analyzer® equipment and makes practicable an automated method for total cyanide. A system for monitoring cyanide in waste water is available.[28] The manifold of this system is shown in Fig. 4.5.

The usual way to carry out distillation in a continuous-flow system is to pass the solution through a coil in a heating bath and then allow the vapour to 'flash' off in a distillation head. In this work on cyanide distillation it was found that better recovery of low levels of cyanide was obtained by using a thin-film distillation system. In this system the sample is passed as a thin film down the inside of a sloping heated tube. The HCN distills rapidly from this film enabling recovery of levels at fractional $\mu g \, l^{-1}$. This technique has been improved by Kelada *et al.*[29] to give an automated system which has a detection limit of $0.25 \, \mu g \, l^{-1}$ CN. In this system the complex cyanides are decomposed by irradiation with ultraviolet light. The total cyanide is vaporized by distillation in a thin-film evaporator, collected in 0.02 M NaOH and determined colorimetrically with chloramine T and pyridine–barbituric acid. The 'simple' cyanides determined this way show good correlation with the 'cyanides amenable to chlorination' determined by the method in *Standard Methods*.[2] Sediments can be analysed in the system by pumping them as a slurry.

In a study of the interferences it was found that the interference from sulfide can be overcome by optimizing the concentration of chloramine T. Thiocyanate is broken down in the ultraviolet irradiation and produces an equimolar amount of cyanide.

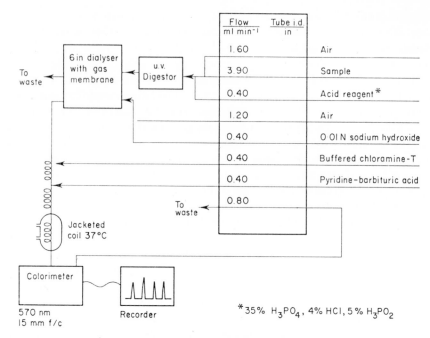

Flow ml min⁻¹	Tube i.d. in	
1.60		Air
3.90		Sample
0.40		Acid reagent*
1.20		Air
0.40		0.01 N sodium hydroxide
0.40		Buffered chloramine-T
0.40		Pyridine–barbituric acid
0.80		

To waste

6 in dialyser with gas membrane

u.v. Digestor

To waste

Jacketed coil 37°C

Colorimeter
570 nm
15 mm f/c

Recorder

*35% H_3PO_4, 4% HCl, 5% H_3PO_2

Fig. 4.5. Manifold used in cyanide monitor.

A potentiometric method has been described[30] which carries out the measurement on the sample without distillation. Any sulfide present is first removed by precipitation. Correction for interference from halides is made by carrying out the determination at pH 2–3, where only the halides are measured, and then at pH 11.5, where both cyanide and halide are measured. Distinction is made between simple and complex cyanides by ultraviolet irradiation. In this case sulfide is added before the irradiation step to precipitate the metals liberated by the breakdown of the complex cyanides and drive the reaction forward. Excess sulfide is then removed before the potentiometric determination. The detection limit with this technique is 2 μg l⁻¹ CN.

An alternative method for the measurement of the molecular hydrogen cyanide in water is to extract the sample with a solvent. In the method described by Montgomery et al.[31] the water sample was extracted with 1,1,1-trichlorethane. This removed a small fraction of the molecular hydrogen cyanide in the sample. The hydrogen cyanide was back-extracted into a solution of sodium pyrophosphate and determined colorimetrically. The method measures HCN down to about 10 μg l⁻¹ HCN.

Water samples

As described above, when the water sample is taken, oxidizing agents are tested for with *o*-tolidine reagent. If they are present they should be removed with ascorbic acid. If sulfides are present they should be precipitated as cadmium sulfide and then the sample should be brought to a pH above 12 with sodium hydroxide.

Sediment samples

Samples of sediment are extracted with water to determine the soluble cyanide level. For a total cyanide determination of a solid sample the sample can be placed in the distillation flask and distilled as described above. To solubilize insoluble cyanides the solid sample should be leached with a 10% NaOH solution for 12–16 h before the distillation is carried out.

Nitrogen forms

The nitrogen forms that are distinguishable in solutions are nitrite, nitrate, ammonia, 'total Kjeldahl nitrogen' and 'total nitrogen'.

Nitrite and oxides of nitrogen in air

Nitrite in solution is determined by allowing the nitrite at a pH of 2.0–2.5 to diazotize sulphanilamide and then coupling this diazo salt with N-1-naphthyl-ethylenediamine to form an azo dye. The concentration of the pink dye is determined by measuring the absorbance at 543 nm. This is a sensitive analytical method; the detection limit in the manual method is 1 μg l^{-1} N. The method may be carried out in Auto Analyzer® equipment.

Nitrogen dioxide in air may be measured by the same colour reaction. In a typical manual determination, the air is sampled through a fritted bubbler containing a mixture of sulfuric acid and N-1-naphthylethylenediamine (Griess–Saltzmann reagent). The air is sampled for about 30 min and the colour allowed to develop for at least a further 15 min, then measured spectrophotometrically. The measurement may be calibrated using standard solutions of sodium nitrite. However, it is best calibrated using standard gas concentrations of NO_2 since there is some controversy as to the stoichiometry of the conversion factor of nitrite to NO_2. This method will measure 40–1500 μg m^{-3} NO_2 in air.

The azo dye produced with the Griess–Saltzmann reagent is somewhat unstable, so that sampling times should be less than 1 h and the colour should be determined within 4 h of sampling. Hence when sampling for longer periods, such as the 24-h period used as the basis for some regulatory purposes, the Jacobs–Hochheiser[32] procedure or some modification of it is used. In this the air sample is passed through sodium hydroxide and the NO_2 reacts to form sodium nitrite. This is then measured by the colorimetric procedure above. In a study of the 24-h manual method for a determination of NO_2 Mulik et al.[33] showed that this collection in NaOH had deficiencies such as variable collection efficiency and interferences from nitric oxide, formaldehyde, carbon monoxide and ammonia. They recommend a solution containing triethanolamine and guaiacol to collect the NO_2, with the addition of sodium metabisulphate to preserve the nitrite if the sample cannot be analysed within 48 h. Alternative colorimetric procedures were developed to analyse the solution.

The nitrite produced by the absorption of NO_2 into solution can also be measured using the nitrogen oxide electrode.[15a,34] The determination of nitric oxide (NO) in air is made by oxidizing the NO to NO_2, by bubbling the air through an oxidizing solution such as potassium permanganate. The total NO_2 is then measured and the result compared with the NO_2 in an unoxidized sample, the NO representing the difference.

In more recent techniques of measuring NO_2 in air it is converted to NO and the chemiluminescent reaction with ozone used to measure its concentration. This is further discussed in Chapter 8.

Nitrate

There are a variety of methods available for measuring nitrate in solutions. These are, in order of decreasing sensitivity, reduction of nitrate to nitrite and measurement of the nitrite colorimetrically as above; colorimetric determination with brucine sulfate; colorimetric determination with chromotropic acid; and reduction of the nitrate to ammonia which is determined titrimetrically or colorimetrically after distillation. Nitrate can also be determined by the nitrate ion-selective electrode[15a] and, in some waters, by measuring the absorbance at 220 and 275 nm.

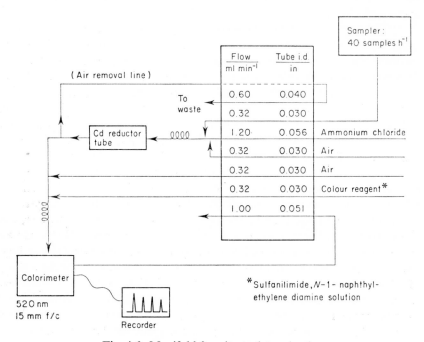

Fig. 4.6. Manifold for nitrate determination.

To carry out the conversion of nitrate to nitrite, reduction with cadmium is used. The manual method uses a column filled with amalgamated cadmium filings. The pH of the sample is adjusted to be below 9, if necessary, and ammonium chloride is added to the sample. After passing through the column the nitrite produced is determined colorimetrically as above. A correction is made for the nitrite originally present in the sample. In the automated method described by Brewer and Riley[35] the sample is mixed with EDTA solution and passed through a coil containing the cadmium filings. In some versions the cadmium filings are coated with copper. The reduction of nitrate to nitrite proceeds with 80–90% efficiency. The EDTA is used for two reasons: the sequestering ability of the EDTA keeps the cadmium surface active and its buffering action enables the reduction to proceed at the optimum pH. The detection limit of the method is $1 \, \mu g \, l^{-1}$ N.

In later versions of the procedure,[36] the EDTA is replaced by ammonium chloride and the sample is diluted with the ammonium chloride–ammonia solution in order to control the pH. The manifold for this determination is shown in Fig. 4.6.

This automated method for nitrate is a sensitive, convenient procedure and is probably the most commonly used method for nitrate determination. However, one difficulty in analysing nitrate is that in environmental samples it may occur over a wide range of levels. This wide range of occurrence is one reason why there are several recommended methods for nitrate, each one covering a different range of levels. It is possible to dilute samples that contain a high level of nitrate in order to analyse them by a sensitive method, such as the cadmium reduction method, but carrying out manual dilutions on a large number of samples becomes a major undertaking. To overcome this difficulty, a system for nitrate determination using the Auto Analyzer® has been described[37] in which progressive dilutions of the sample up to 1000:1 can be made automatically, depending at which point in the manifold the sample is entered. In this way the one manifold can handle the whole range of nitrate levels that might occur.

Nitrate in the range of 0.1–$2 \, mg \, l^{-1}$ N can be determined by a colorimetric method using brucine. The sample is reacted with brucine sulfate in acid solution at 95 °C to produce a yellow colour. The absorbance of this colour is measured at 410 nm. Interferences in the reaction are strong oxidizing agents such as chlorine, strong reducing agents such as nitrite, and chloride. The sample is tested for oxidizing agents using the o-tolidine reagent and, if they are found, sodium arsenite is added. The potential interference from nitrite is overcome by the addition of sulfamic acid to the brucine sulfate reagent. Interference from chloride is swamped by the addition of sodium chloride. Some metals such as iron and manganese will produce a colour in the method, but in normal water samples these do not constitute a significant interference.

An alternative colorimetric method for nitrate in the range 0.1–$5 \, mg \, l^{-1}$ N is that using chromotropic acid. Nitrate reacts with chromotropic acid (4,5-dihydroxy-2,7-naphthalene disulfonic acid) in 80% sulphuric acid to form a yellow colour whose absorbance is measured at 410 nm. Other materials which will also yield yellow colours are residual chlorine and other oxidizing agents, nitrite and chloride. To overcome these interferences in the test method a sodium sulfite–urea reagent is first added. The sulfite removes any oxidizing agents and the urea reacts with any nitrite to form nitrogen gas. Antimony sulfate is added to make the effect of chloride.

In samples containing high levels of nitrate, i.e. >2 mg l^{-1} N, the nitrate can be converted to ammonia by reduction with Devarda's alloy. This is carried out in the Kjeldahl type of distillation equipment. If the ammonia has not been previously stripped from the solution for measurement, the sample is placed in the distillation flask with a sodium hydroxide–sodium borate buffer and the pH adjusted to 9.5 with sodium hydroxide. The flask is boiled and about half of the solution distilled to remove ammonia. The Devarda's alloy is added and the ammonia produced distilled off and collected in a boric acid solution. The ammonia is then determined by titration with a strong acid or by a colorimetric method such as Nesslerization. This determination measures both the nitrite and nitrate in the sample.

The nitrate ion-selective electrode[15a] can be used to measure levels down to about 0.2 mg l^{-1} N. The only significant interferences in natural water samples are chloride and bicarbonate. The chloride ion is removed by precipitating it with silver sulfate. The bicarbonate is removed by adjusting the pH to 4–4.5 with sulfuric acid.

The nitrate ion absorbs in the ultraviolet region of the spectrum and, in relatively clean water, an estimate of the nitrate concentration can be made by measuring the absorbance of an acidified sample at 220 nm. A correction for the presence of organic materials which would also absorb at this wavelength is made by making another measurement at 275 nm. The organic materials which would absorb at 220 nm will also show some absorbance at 275 nm, whereas the nitrate will not.

A description of the methods for nitrate–nitrogen analysis is given in *Standard Methods*.[2]

Nitrate in air particulates may be determined by extraction of the solids collected on a filter or in an impinger with water and analysis of the solution by the methods above. A method for the measurement of the particulate atmospheric nitrate using a flowthrough nitrate electrode unit has been described by Forney and McCoy.[38]

Ammonia

Ammonia in 'clean' waters may be determined by the Nessler method.[2] In this the reaction between ammonia and an alkaline solution of mercuric iodate and potassium iodide produces a yellow to brown colour. The hue of this colour changes from yellow to reddish brown as the amount of ammonia nitrogen increases, and the absorbance measurements are made over the range 400–500 nm depending on the concentrations being measured. To prevent precipitation of calcium and magnesium salts under the alkaline conditions of the test EDTA or Rochelle salt is added. For samples which are other than 'clean' the ammonia is best separated by distillation before the colour-forming step. The distillation may be carried out in a Kjeldahl type of distillation equipment; the sample is brought to a pH of 9.5 with sodium borate and sodium hydroxide, and the ammonia distilled off is collected in a solution of boric acid. The method will measure down to 20 μg l^{-1} ammonia nitrogen.

When the ammonia is at high levels, i.e. 5 mg l^{-1} N, the distillate above may be determined by titration with a strong acid.

Ammonia in solution may also be determined by the Berthelot reaction. In this reaction a blue compound, indophenol is formed when ammonia, hypochlorite and phenol react. The colour is measured at 630 nm. Catalysts may be used to increase the intensity of the colour; common catalysts are manganous salts, sodium nitroprusside or acetone. There is interference in the reaction by metal ions. This interference may be suppressed by the addition of a complexing agent such as citrate, tartrate or EDTA.

A discussion of the factors affecting the Berthelot reaction is given by Fleck.[39] A description of the procedure for the determination of ammonia, by both a manual and automated procedure using the Berthelot reaction is given in Standard Methods.[2] The detection limit by these methods is 10 μg l^{-1} N.

In many natural waters the detection limit of 10 μg l^{-1} N is not low enough. In some waters which are relatively unpolluted, levels down to 1 μg l^{-1} N are found. The addition of the complexing agent in the Berthelot reaction decreases the sensitivity and for these relatively 'clean' waters it is possible to decrease the level of complexing agent used, since there are generally low levels of interfering metals present. In this way a detection limit of less than 1 μg l^{-1} N is obtainable. A typical manifold for 'clean' natural waters is shown in Fig. 4.7.

For waters containing high levels of interfering materials and low levels of nitrogen, the automated method of Sawyer and Grisley[40] may be used. In this method the ammonia in the sample is dialysed into a clean reagent stream where it is treated with sodium hypochlorite to form chloramine. The excess hypochlorite is destroyed and the chloramine is reacted with o-tolidine. The absorbance of the resultant dye is measured at 420 nm. The detection limit with this method is 1 μg l^{-1} N.

One of the problems in measuring low levels of ammonia is that often the samples can absorb ammonia from the laboratory atmosphere as they sit in the tray of the sampler. Some samplers have a cover which fits over the tray and this can be used to avoid such contamination. An alternative which has been found effective is to cover the sample cups in the tray with 'saran wrap'. The sampler probe is sharpened so that it pricks a hole in the saran to take the sample.

An alternative procedure for ammonia levels in the 100 μg l^{-1} N range and above is to use the ammonia electrode. In the procedure described by Evans and Partridge,[15a,41] using an electrode with a gas-permeable membrane, the sample was mixed with a pH adjusting solution containing sodium hydroxide and EDTA

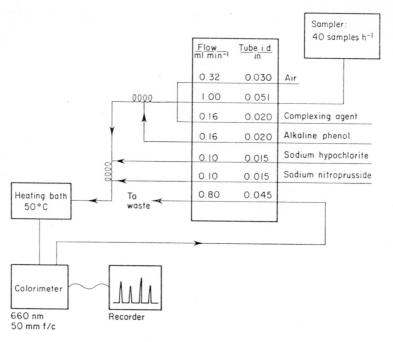

Fig. 4.7. Manifold for ammonia determination in 'clean' water.

and the potential measured. Residual chlorine in the water was destroyed with sodium thiosulphate. The detection limit was 30 μg l^{-1} N, with a suggested practical lower limit of 100 μg l^{-1} N.

Kjeldahl nitrogen

In this determination the sample is digested with sulfuric acid. The amino nitrogen of most organic compounds is converted to ammonium sulfate. This ammonium sulfate is then measured. If the ammonia nitrogen is removed (by distillation) before the digestion, or a correction is made for its presence, the Kjeldahl nitrogen determination gives a measure of the organic nitrogen in the sample. If the ammonia is included in the determination, the result is expressed as 'total Kjeldahl nitrogen'. The method does not measure nitrogen present as azide, azine, azo, hydrazone, nitrate, nitrite, nitro, nitroso, oxime, or semicarbazone.

The determination consists of two parts—digestion and measurement of the ammonia formed. There are many variations, such as a manual digestion followed by neutralization and distillation of the ammonia from the reaction mix for determination by titration or other methods. Alternatively the ammonia may be determined in the digestion mixture, after dilution and neutralization, by colorimetric methods such as the Berthelot reaction or with an ammonia electrode. The manual digestion may be followed by an automated determination of the ammonia, or the whole digestion–ammonia measurement may be carried out in an automated system. The manual digestion may be carried out in 'full-size' Kjeldahl flasks with a total capacity of about 800 ml, or a variety of 'micro-Kjeldahl' flasks and heating systems are available. Alternatively the digestion can be carried out in sealed tubes.

In a typical manual digestion procedure the sample, which may be a water sample, a sediment or air particulate sample, is placed in the flask, with water if necessary and the sodium hydroxide added. The flask is boiled to remove ammonia (if this is desired) which may be collected and determined. The flask is cooled and a digestion mixture of sulfuric acid, potassium sulfate and a catalyst is added. The catalyst used most often is mercuric sulfate; selenium dioxide and copper oxide are also used. Oxidizing agents such as perchlorate are sometimes added. When mercury is used as a catalyst it forms a mercury–ammonium complex that must be decomposed before the ammonia is measured. Sodium thiosulfate or zinc dust are used to carry out this decomposition. After the digestion is complete the flask is cooled, the mercury–ammonium complex is decomposed and the ammonium salts produced determined by any of the methods described above for the determination of ammonia. The most

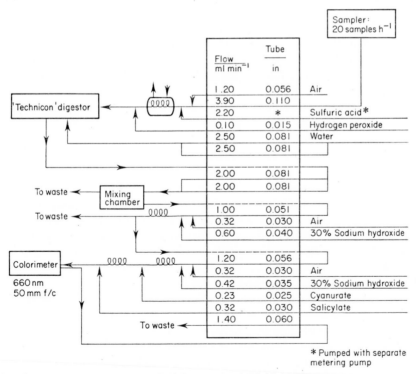

Flow ml min⁻¹	Tube in	
1.20	0.056	Air
3.90	0.110	
2.20	*	Sulfuric acid*
0.10	0.015	Hydrogen peroxide
2.50	0.081	Water
2.50	0.081	
2.00	0.081	
2.00	0.081	
1.00	0.051	
0.32	0.030	Air
0.60	0.040	30% Sodium hydroxide
1.20	0.056	
0.32	0.030	Air
0.42	0.035	30% Sodium hydroxide
0.23	0.025	Cyanurate
0.32	0.030	Salicylate
1.40	0.060	

Sampler: 20 samples h⁻¹

* Pumped with separate metering pump

Fig. 4.8. Manifold for low-level Kjeldahl nitrogen.

commonly used method appears to be the Berthelot reaction. This may be carried out in Auto Analyzer® equipment.

The sensitivity of the Berthelot reaction is much reduced by the high salt content of the reaction mix from the Kjeldahl digestion and by the complexing agents added to overcome interference from metals. An alternative procedure which is about 30 times more sensitive is the reaction between ammonia and a weakly alkaline mixture of sodium salicylate and dichloroisocyanurate.[42]

Traditionally distillation of the reaction mix and determination of the ammonia by the Nessler method has often been used after Kjeldahl digestion. However, this seems to have been almost completely displaced by the Berthelot reaction and others.

In the automated digestion the sample and the digestion mix are pumped into the end of a glass helix. This helix is rotated at a constant speed and the liquid moves along the helix. The helix is heated and the evaporation of part of the liquid results in a thin film being formed which slowly moves over the heated glass surface. The conditions used in this technique are somewhat different from the manual digestion. Sulfuric acid alone is often used rather than a potassium sulfate–sulfuric acid mixture because the mixture can give some problems in the resampling procedure after digestion. Oxidizing agents such as hydrogen peroxide and perchloric acid are commonly used. The thin film formed in the helix makes it easy to lose nitrogen by volatilization of ammonium sulfate, so that good temperature control and calibration with organic nitrogen standards are necessary. A method for determining low levels of Kjeldahl nitrogen in water is described by Elkei.[43] (This uses the sodium salicylate–dichloroisocyanurate colour reaction.) The manifold for this method is shown in Fig. 4.8. The detection limit is $10 \, \mu g \, l^{-1}$ N.

A discussion of the various methods of determining organic nitrogen is given by Fleck.[39]

Total nitrogen

Because nitrogen is a nutrient for plant growth, there is often conversion between the various nitrogen forms after the sample has been taken. In many cases a satisfactory method of measuring nitrogen pollution in water (e.g. that from sewage effluents) is to determine the total nitrogen content rather than be concerned with the different nitrogen forms. A convenient way to measure total nitrogen is to convert all the nitrogen compounds to one form and then determine it. Two methods which are in common use convert all the nitrogen to nitrate, one method using ultraviolet irradiation, the other using an alkaline persulfate digestion.

The ultraviolet irradiation may be carried out on a manual basis, followed by automated determination of the nitrate formed, as in the method of Henriksen.[44] A fully automated method has been described.[45] In this method the irradiation is carried out by passing the sample through a silica coil around the outside of a mercury lamp. The oxidation of nitrogen compounds was found to be pH-dependent. Urea oxidizes faster under acid conditions; ammonium salts oxidize faster under alkaline conditions. Because of this the irradiation is carried out in two stages—one at pH 2 and one at pH 8.

An alternative procedure is to oxidize with peroxydisulfate.[46] This reaction is carried out under alkaline conditions at 100 °C in an autoclave at 30 psig.

Total nitrogen levels may also be determined by combustion techniques. In a method using the Microcoulometer® as a detector,[47] the water samples are injected into a furnace where the samples are pyrolysed, and the nitrogen compounds reduced to ammonia. The ammonia produced is then measured in the microcoulometer. In another method the samples are injected into a combustion furnace where oxides of nitrogen are produced. These are then measured by a chemiluminescent reaction with ozone.[48]

Water samples

The various nitrogen forms are used as indicators of types of pollution. For example, levels of ammonia in a natural water give indication of pollution by sewage. The nitrogen forms are active in the biological processes that occur in natural waters and may change drastically after the sample is taken unless they are preserved.

All of the nitrogen forms except nitrite may be preserved by the addition of sulfuric acid (0.8 ml l^{-1} H_2SO_4). Nitrite is best preserved by cooling the sample to about 4 °C and analysing on the same day. If this is not possible, 40 mg l^{-1} $HgCl_2$ may be added as a preservative, but the sample must still be stored at low temperature. When samples are to be analysed for ammonia and there is residual chlorine in the water, this should be destroyed with sodium sulfite before the acid preservative is added.

Soil–sediment samples

The nitrogen forms in soils and sediments are extracted with a variety of aqueous solvents, and these solutions are analysed by the methods described above. In many cases the extractant solution can be used to overcome interferences in the analytical method. For example, in the determination of nitrite and nitrate in soil extracts, there is interference from chloride, coloured organic materials and colloidal material. The recommended extractant is a solution of $CuSO_4$–$AgSO_4$, since this will precipitate chloride and will coagulate the colloidal and organic material. The solution normally used for ammonia extraction is 1 M or 2 M potassium chloride. This will extract the 'exchangeable' ammonium ions from the soil or sediment. A total nitrogen determination can be carried out on the solid itself; the ammonia produced is separated by subsequent distillation.

Phosphorus

The phosphorus forms that are usually measured in environmental samples are orthophosphate, acid-hydrolysable phosphate and total phosphorus. The acid-hydrolysable phosphate comprises the orthophosphate plus the condensed phosphates. The total phosphorus consists of the inorganic phosphates and the organically bound phosphorus. Phosphorus is a nutrient that is essential to life and a large part of the development of its analytical methodology in water is concerned with the effect that it has on plant life. In many fresh waters it has been demonstrated that phosphorus is the nutrient whose availability is the limiting factor in plant growth and hence the determining factor in the extent to which eutrophication takes place.

Orthophosphates are the salts of orthophosphoric acid, H_3PO_4. The condensed phosphates (or polyphosphates) are formed by the condensation of the ortho-phosphate with elimination of water to form polymers containing two units (pyrophosphate), three units (tripolyphosphate) and higher (e.g. hexameta-phosphate). These polymers will hydrolyse and 'revert' to orthophosphate under suitable conditions.

Most of the methods used for the determination of phosphorus in environmental samples involve the formation and measurement of molybdenum blue. This determines orthophosphate; the other forms of phosphorus are converted to orthophosphate for measurement.

Orthophosphate

Orthophosphates react with ammonium molybdate in an acid solution to form a yellow heteropoly molybdophosphoric acid. This is then reduced to molybdenum blue with a reducing agent. The absorbance is measured at 660 nm or 880 nm; at 880 nm the extinction coefficient is about three times that at 660 nm.

There are many procedures that can be used to carry out this measurement, involving different acid media (generally hydrochloric acid or sulfuric acid) and a variety of reducing agents. The sensitivity of the method is dependent on the reducing agent; ones which are commonly used are 1-amino-2-naphthol-4-sulfonic acid (ANSA), ascorbic acid, and stannous chloride.[2] When using ANSA, bismuth is sometimes added to increase sensitivity; with ascorbic acid, antimonyl potassium tartrate is usually added since the antimony phospho-molybdate complex formed is rapidly reduced with ascorbic acid. Using manual procedures the lowest detectable levels of the three methods are: ANSA, 300 μg l^{-1} P; ascorbic acid, 10 μg l^{-1} P; stannous chloride, 3 μg l^{-1} P. When levels lower than about 10 μg l^{-1} are to be determined the molybdenum blue colour is best measured after extraction into a solvent.[2]

When an automated procedure is used the detection limits with ascorbic acid and with stannous chloride are about 1 μg l^{-1} P. A typical manifold used on the Auto Analyzer® for the stannous chloride method is shown in Fig. 4.9. In the automated method described by Henriksen,[49] the yellow heteropoly acid is extracted into a solvent and then reduced with stannous chloride.

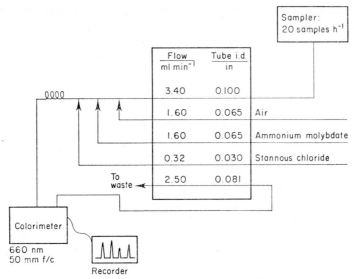

Flow ml min^{-1}	Tube i.d. in	
3.40	0.100	
1.60	0.065	Air
1.60	0.065	Ammonium molybdate
0.32	0.030	Stannous chloride
2.50	0.081	

Fig. 4.9. Manifold for orthophosphate determination.

Two potential interferences in the method are silicate and arsenate. These both form similar heteropoly acids which are reduced to form a blue colour. The interference from silicate can be overcome by controlling the acid concentration in the procedure at a level at which silicate does not form the complex. In fact by merely changing the acid concentration the same method can be used to determine phosphate or silicate. To overcome the interference from arsenate it is reduced to arsenite. In the procedure described by Johnson[50] the reduction is carried out with thiosulfate in an acidic solution. A large excess of sodium metabisulfite is added to prevent the formation of colloidal sulfur. An automated method for removing the arsenate interference has been described.[51] One important aspect of this method of overcoming the arsenate interference is that the reaction can be carried out at room temperature at low acid concentration, thus minimizing the likelihood of hydrolysing other phosphorus forms to orthophosphate.

An alternative method of determining orthophosphate is by pulse polarography.[52] The molybdenum blue is extracted into isoamyl alcohol and then back-extracted into a tartrate buffer and polarographed. With the five-fold concentration achieved in the extraction a detection limit of about 0.6 µg l^{-1} P is obtained.

If the ammonium molybdate and orthophosphate are reacted in the presence of vanadium, the yellow vanadomolybdophosphoric complex is formed.[2] This may be used to determine orthophosphate by measuring the absorbance at 470 nm. The detection limit is about 100 µg l^{-1} P. Orthophosphate also reacts with sodium tungstate to form a similar compound which can be used to determine phosphorus.

It is not universally accepted that the molybdenum blue determination measures only the phosphorus present as orthophosphate. Rigler[53] has suggested that part of the phosphorus measured in lake water does in fact come from the hydrolysing of organic phosphorous compounds.

Acid-hydrolysable phosphate

In this determination the sample is treated with acid to hydrolyse the condensed phosphates such as pyro-, tripoly- and hexametaphosphate to orthophosphate. Besides these condensed inorganic phosphates some organic phosphate compounds are also hydrolysed to orthophosphate. The method has been described as the determination of inorganic phosphate but it is better described as the determination of the acid-hydrolysable phosphate. After the hydrolysis the orthophosphate is determined by one of the methods described above.

The hydrolysis is either carried out by making the sample 0.1 N with sulfuric acid and boiling for 90 min or by autoclaving the acidified sample at 15–20 psig for 30 min. After this hydrolysis the sample is cooled, brought back to volume if necessary and analysed for orthophosphate. To maintain the optimum acid concentration in the colour forming step, adjustment of the acid concentration in the ammonium molybdate reagent is made to compensate for the acid added in the hydrolysis.

Total phosphorus

In this determination the sample is digested under conditions in which organic materials are oxidized and the organic phosphorus compounds are broken down to convert the phosphorus to orthophosphate. The condensed phosphates are also hydrolysed to orthophosphate in this process. The total amount of phosphorus present is then determined as orthophosphate. The rigour of the digestion procedure depends on the type of sample being analysed. The three procedures described in *Standard Methods*[2] for water samples in order of decreasing rigour are, digestion with perchloric acid, with a sulfuric–nitric mixture or with acid peroxydisulfate. The peroxydisulfate digestion is carried out by boiling on a hot plate for 30–40 min or by autoclaving for 30 min at 15–20 psig. Other methods that have been used are a digestion with 50% hydrogen peroxide and fusion with magnesium nitrate. The peroxydisulfate digestion is the one recommended in ASTM Standards[54]; it appears to be the one most used, perhaps because it is most convenient. The particular advantage of the procedure using the autoclave is that the orthophosphate determination is sensitive to the acid concentration, and there is no significant change produced in this in the autoclave. The digestion procedures using a hot plate involve the sample being evaporated to small volume. At this stage more or less acid fumes are lost and it is difficult to be sure that the subsequent dilution and acid concentration adjustment results in a uniform acid concentration in the orthophosphate determination step.

The Auto Analyzer® continuous digestion has been used for a fully automated determination of total phosphorus in water and sewage sludge. Tyler and Biles[55] used a digestion mix of sulfuric acid, perchloric acid and a vanadium pentoxide catalyst to convert all the phosphorus forms to orthophosphate. Milbury et al.[56] used the same type of digestion mix in the simultaneous determination of nitrogen and phosphorus in sewage sludge using the continuous digestion.

An alternative to chemical digestion for the breakdown of organophosphorus compounds is the use of ultraviolet irradiation. This was used by Armstrong and Tibbitts,[57] in a manual procedure, to measure phosphorus, nitrogen and carbon in sea water. The ultraviolet irradiation breaks down the organophosphorus compounds but does not hydrolyse inorganic polyphosphates unless the temperature in the irradiated sample is allowed to rise. This method offers the possibility of measuring organic phosphorus separately from a total phosphorus

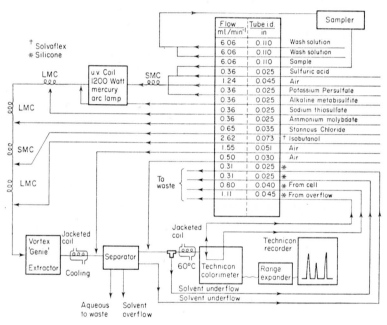

Flow ml/min⁻¹	Tube i.d. in	
6.06	0.110	Wash solution
6.06	0.110	Wash solution
6.06	0.110	Sample
0.36	0.025	Sulfuric acid
1.24	0.045	Air
0.36	0.025	Potassium Persulfate
0.36	0.025	Alkaline metabisulfite
0.36	0.025	Sodium thiosulfate
0.36	0.025	Ammonium molybdate
0.65	0.035	Stannous Chloride
2.62	0.073	† Isobutanol
1.55	0.051	Air
0.50	0.030	Air
0.31	0.025	*
0.31	0.025	*
0.80	0.040	* From cell
1.11	0.045	* From overflow

Fig. 4.10. Total phosphorus determination.

determination. Henriksen[44] used a manual irradiation procedure to measure nitrogen, phosphorus and iron in fresh water. He found it necessary to add sulfuric acid to the sample to overcome interference from iron in the phosphorus determination. Grasshoff[58] used an automated system for irradiating samples in the determination of total phosphorus in sea water. To hydrolyse the inorganic polyphosphates the samples were boiled before the irradiation. An automated procedure has been reported[59] which used ultraviolet irradiation followed by reduction with thiosulfate to remove arsenate interference, development of the molybdenum blue with stannous chloride and extraction into isobutanol. The temperature in the ultraviolet irradiation stage is allowed to rise to hydrolyse the polyphosphates. The manifold used is shown in Fig. 4.10. The detection limit is 0.2 μg l^{-1} P. In this work natural water samples from the Great Lakes were analysed by the automated ultraviolet procedure and by a procedure where the ultraviolet irradiation was replaced by manually autoclaving the samples with peroxydisulfate. It was found that the ultraviolet irradiation yielded results that averaged 97% of the values obtained with the autoclave.

In an evaluation of the procedures for determining low levels of phosphorus in fresh water MacKay[60] concludes that the preferred procedure to measure phosphorus in a manual procedure with a detection limit of 1 μg l^{-1} P, was to extract the molybdenum blue into isobutanol. In this work the oxidation was carried out on a hot plate with persulfate, the arsenate was reduced to arsenite and the molybdenum blue colour was developed with ascorbic acid–antimonyl tartrate reagent.

Condensed phosphates

The various condensed phosphates are determined by chromatographic procedures. In a procedure reported by Scott and Haight,[61] ortho-, pyro- and tripolyphosphate were separated by anion exchange thin layer chromatography. The solution was spotted on a plate coated with cellulose impregnated with polyethyleneimine and the separation carried out by ascending chromatography using 2 M lithium chloride as the solvent. After the separation (which took 2 h) the plates were sprayed with ammonium molybdate and ascorbic acid and incubated at 100 °C. After about 5 min hydrolysis of the polyphosphate spots was complete and the molybdenum blue colour developed.

Yoza et al.[62] have used a gel chromatographic column (Sephadex G-25) to separate condensed phosphates. They added magnesium to the phosphate solution and the various magnesium phosphate compounds were eluted from the column at different times. An atomic absorption spectrometer was used to detect the magnesium complexes of the condensed phosphates as they were eluted.

Heinke and Behmann[63] have used gradient elution ion exchange chromatography for the separation and determination of ortho-, pyro- and tripolyphosphate in waste water and lake water. The phosphates were adsorbed on an ion exchange column and then eluted with potassium chloride solution using Auto Analyzer® equipment. The eluant from the column was passed through a heating bath to hydrolyse the polyphosphates and the resulting orthophosphate determined as molybdenum blue.

Water samples

Orthophosphate is believed to be the form in which phosphate is readily available for plant growth and thus, in studying a phenomenon such as the effect of phosphorus pollution on algal blooms, it is the level of orthophosphate which is believed to be of the greatest significance. Hence very often water samples are taken for the purpose of measuring orthophosphate. There is, however, no satisfactory chemical preservative that will stop the biological action in the sample without at the same time causing changes in the distribution of the phosphorus forms. It has been recommended[2] that 40 mg of mercuric chloride per litre be added to samples to preserve them but this causes precipitation of mercurous chloride during the reduction of the heteropoly acid. The most satisfactory way to treat a sample in which orthophosphate is to be determined is to filter immediately, store at low temperature, i.e. 4 °C, and analyse within 24 hours. Samples for inorganic phosphate should be stored at low temperature and also analysed promptly. Samples for 'total' phosphate may be preserved by the addition of 0.2% sulfuric acid. Glass, rather than plastic, bottles should be used for samples containing low levels of phosphorus because phosphate may be adsorbed on the walls of plastic bottles. A convenient way to collect samples for total phosphorus analysis is to use a small (100-ml) borosilicate bottle, and fill it to the pre-determined mark. Acid and peroxydisulfate can be added to the bottle in the laboratory and the bottle autoclaved.

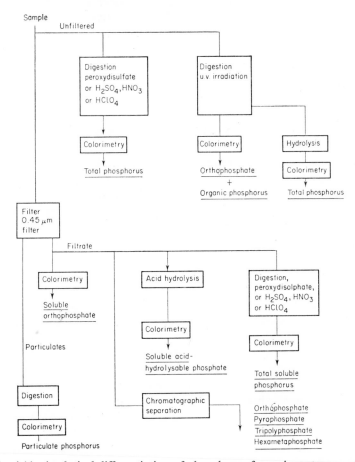

Fig. 4.11. Analytical differentiation of phosphorus forms in water sample.

In many analyses it is often the practice to use an unfiltered sample to measure the 'total' level of the parameter being measured. This is a potentially dangerous practice in phosphorus analysis. The extinction coefficient of molybdenum blue is low and any turbidity in the sample can contribute a significant portion of the optical absorbance measured. In addition, the molybdenum blue does not form a solution in water but is a fine suspension. The particle size affects the optical absorbance. If this suspension is formed in the presence of particulates, it may precipitate on these particles and yield a false absorbance reading. If a 'total' determination is required the heteropoly acid solution should be filtered before the reduction to molybdenum blue, or the molybdenum blue should be extracted into a solvent.

When the level of particulates in water becomes significant the digestion procedure used for a total phosphorus determination may not release all the phosphorus from these particulates. When this is suspected the particulates should be separated and examined by the techniques used for sediments and soils. Fuhs[64] has analysed particulate biological material by dissolving it in 8 N alkali before carrying out a persulfate oxidation.

An analytical scheme for differentiating the phosphorus forms in water is shown in Fig. 4.11.

Sediments and soils

There is often interest in differentiating the phosphorus forms in soil and sediment samples because the form of the phosphorus may give some indication of its origin and also of its availability as a nutrient. Schemes for the fractionation of the inorganic phosphorus have been developed. In the one by Chang and Jackson[65] and later variations the sample is sequentially extracted by ammonium chloride, ammonium fluoride, sodium hydroxide and hydrochloric acid. This separates the inorganic phosphates in the interstitial water and the weakly sorbed phosphate ions from the 'aluminium phosphate', the 'iron phosphate' and the 'calcium phosphate'. However, there has been criticism of these procedures, particularly when they are used for sediment analysis, because of resorption of ions released by the chemical treatment in one stage which may then be extracted in the subsequent treatment. A system used by Williams et al.[66] for sediments from Lake Erie classifies the inorganic phosphorus into two groups, 'apatite-phosphorus' and 'non-apatite inorganic phosphorus'. This distinction is made by extracting the sediment with a mixture of sodium citrate, sodium bicarbonate and sodium dithionite solution (CDB reagent) and then with sodium hydroxide solution. This extracts the non-apatite inorganic phosphorus. The sample is next treated with hydrochloric acid which extracts the apatite-phosphorus. This is a scheme similar to one proposed for soils;[67] the non-apatite inorganic phosphorus can be sub-divided further if required. The organic phosphorus is determined using the extraction method of Mehta et al.[68] The extraction is made with concentrated hydrochloric acid at room temperature, with 0.5 N sodium hydroxide at room temperature, then with 0.5 N sodium hydroxide at 90 °C.

These extracts are combined and the solution analysed for total phosphorus, by a perchloric acid digestion, and for acid-hydrolysable phosphate. The difference represents the organic phosphorus.

Another method for the determination of inorganic, organic and total phosphate in sediments[69] is to extract the sediment with 1 N hydrochloric acid before and after portions of the sediment are ignited at 550 °C. The total phosphorus is that extracted after ignition; the inorganic phosphorus is that extracted without ignition; the organic phosphorus is represented by the difference.

An alternative method for determining the total phosphorus in soils and sediments is to carry out a fusion with sodium carbonate. A discussion of the phosphate chemistry in lake sediment is given by Syers *et al.*[70]

Other samples

Biological samples may be digested for total phosphorus analysis by the perchloric acid digestion or with nitric–sulfuric acid. Peroxydisulfate digestion does not proceed very rapidly with biological materials unless the sulfuric acid concentration is increased to improve the solubility of organic materials. This increase in acid concentration tends to lower the stability of the peroxydisulfate to spontaneous decomposition. Biological samples may also be treated by a Kjeldahl digestion to release their phosphorus.

Air particulates may be treated by the same extraction procedure as is recommended for sediments and soils.

Silicon

Forms of silicon more or less in solution can be determined by atomic spectroscopy, by a gravimetric procedure or by colorimetric methods.[2]

The atomic absorption procedures are described in Chapter 2. Silicon forms a refractory oxide and is determined using a nitrous oxide–acetylene flame. The detection limit is about 20 μg l^{-1} Si.

Where solutions contain at least 20 mg l^{-1} SiO_2 a gravimetric procedure may be used. The silica and silicates are precipitated as SiO_2 and then volatilized as SiF_4. In this procedure the sample is evaporated in a hydrochloric acid solution and the residue dried at 110 °C. This decomposes silicates and precipitates silicic acid which is partially dehydrated. The residue is re-suspended in hydrochloric acid and filtered off. The dehydration may not be complete in the one treatment and a second evaporation and drying is carried out on the filtrate. The residues from the two dehydrations are combined and ignited at 1200 °C. The residue is weighed and then treated with hydrofluoric acid to volatilize the silicon as SiF_4. Reweighing determines the silicon present. An alternative procedure is to use perchloric acid in the dehydration of the precipitated silicic acids. This gives a silica precipitate that is easier to filter than the one obtained with hydrochloric acid. The recovery of silica with one evaporation is also higher with perchloric acid. The volatilization of the precipitated silica as SiF_4 is carried out because the silicic acids tend to coprecipitate other metal ions from the sample. In many cases this does not cause a significant error and the determination of the weight of the ignited silica is sufficient. This gravimetric procedure is used to calibrate the silicate standard solutions used in the colorimetric method below. Two colorimetric procedures may be used, both involving the formation of molybdo-silicic acid. In the first method the colour of the yellow molybdosilicic acid is measured; in the second, more sensitive method a reduction of this acid is carried out to form a heteropoly blue compound. The sample is reacted with ammonium molybdate at a pH of approximately 1.2. Both silicate and phosphate in the sample react to form heteropoly acids. The phosphate heteropoly acids are decomposed by the addition of oxalic acid and the yellow colour of the molybdosilicic acid is measured at 410 nm. A calibration curve using standard silicate solution is prepared. If the colour comparison is to be made visually then permanent colour standards can be prepared using potassium chromate and borax solutions. This method is used for silica levels ranging from 0.4 to 25 mg l^{-1} SiO_2. For determining lower levels of silica the heteropoly acid is reduced using 1-amino-2-naphthol-4-sulfonic acid (ANSA). The absorbance of the resulting heteropoly blue colour is measured at 815 nm. The method will measure levels down to about 20 µg l^{-1} SiO_2.

The heteropoly blue method is used in the Auto Analyzer® to measure levels to 5µg l^{-1} SiO_2.

Water samples

Water samples for silica analysis should be collected in plastic rather than glass bottles to avoid the possibility of contamination. Silica occurs in natural waters as silicic acid and its salts. It is generally found that in water samples only a fraction of the silicon that will pass through a 0.45-µm membrane filter will react in the molybdosilicate reaction. This is described as the 'reactive silica'. There is no general agreement as to what this phenomenon means. It has been postulated that the silicic acids polymerize very easily and that only the short (i.e. up to three or four silicic acid units) straight chain polymers comprise the 'reactive silica'. To measure the total soluble silica the sample is digested with sodium bicarbonate solution at 100 °C for an hour to convert all the silicon present to 'reactive silica'.

Acidification of a sample containing silicates will produce silicic acids which tend to polymerize and dehydrate on aging. Because of this no preservatives are added to a sample to be analysed for silica. Samples containing low levels which are to be analysed for 'reactive silica' should be analysed within eight hours of collection.

Air particulates

These are best determined by treatment with hydrogen fluoride in a Teflon®-lined bomb, followed by determination of the silica by atomic absorption (see Chapter 2).

Sediments and soils

The solid samples are treated by the 'standard' geochemical treatments to solubilize silica, such as alkaline fusion.

Sulfur forms

Sulfate

Sulfate in solution may be determined by titration with barium chloride in an alcoholic solution at pH 3.8–4.0 using Thorin as indicator. Thorin is 2(2-hydroxy-3,6-disulfo-1-naphthylazo)benzene arsonic acid. Other cations interfere with the titration and these are removed by first passing the sample through a

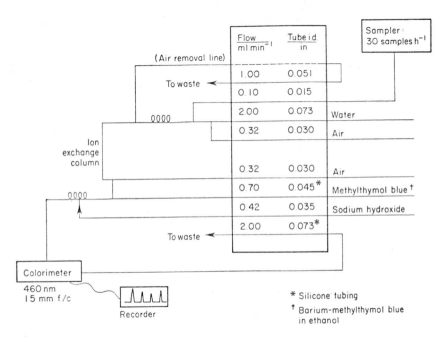

Fig. 4.12. Manifold for sulfate determination.

column of cation exchange resin. Sulfate may also be determined turbidi-metrically as barium sulfate. These procedures are described in ASTM Standards.[54]

An automated procedure is available.[71] In this procedure the sample is passed through a cation exchange column to remove interfering cations and then reacted with barium chloride to form barium sulfate. The unreacted barium chloride reacts with methylthymol blue at a pH of 12.5–13.0 to form a blue complex. The absorbance of the blue complex is measured at 460 nm. The manifold used in this determination is shown in Fig. 4.12. The detection limit is $10 \ \mu g \ l^{-1} \ SO_4$.

In air sampling the sulfates are collected in an impinger or packed mist separator containing distilled water. The sulfate in solution may then be determined by one of the above methods or, more commonly, by the barium chloranilate method. In this method the sample is freed from interfering cations by passage through a cation exchange column and reacted with barium chloranilate.[72] The acid chloranilate released by this reaction is determined by measuring the absorbance at 530 nm. Aerosols collected on filters have been analysed for sulfate and nitrate by ion chromatography.[73]

Sulfur dioxide—sulfite

To measure the concentration of sulfur dioxide in air samples, the gas is absorbed in a solution of potassium tetrachloromercurate to form a dichlorosulfitomercurate complex. In this complex the sulfur dioxide is protected from being oxidized to sulfate by the air blown through the solution. This complex is then reacted with formaldehyde and p-rosaniline to form p-rosanilinemethylsulfonic acid. (This is the modified West and Gaeke procedure.[74]) The absorbance at 560 nm is measured to determine the concentration of sulfur dioxide. Interference from nitrogen dioxide is overcome by the addition of sulfamic acid; EDTA is added to remove the interference from heavy metals. This colour determination may be carried out on the Auto Analyzer®; the manifold used is shown in Fig. 4.13. The detection limit depends on the volume of air sampled, the efficiency of absorption, etc. It is of the order of 20 μg m^{-3} SO$_2$.

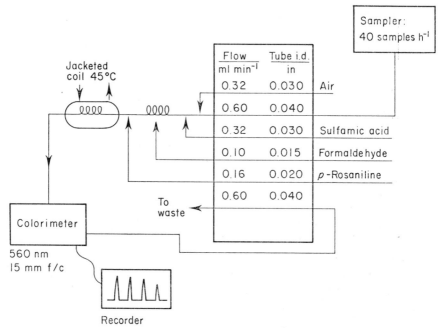

Fig. 4.13. Manifold for determination of SO$_2$ in manually impinged air samples.

An alternative procedure is to absorb the sulfur dioxide in a solution of hydrogen peroxide and determine the resulting sulfate either with barium chloranilate or by a titrimetric method.

Sulfite in water samples is determined by the same procedure used to determine sulfur dioxide in air by the modified West and Gaeke procedure above.[74] The detection limit is 60 $\mu g\,l^{-1}\,SO_3^{2-}$.

Sulfides—hydrogen sulfide

Sulfide in solution may be measured colorimetrically by the formation of methylene blue.[2] The sulfide solution is reacted with an acid solution of N,N-dimethyl-p-phenylenediamine oxalate and ferric chloride. After a few minutes the colour develops and the colour due to the ferric chloride is removed by the addition of an ammonium phosphate solution. Measurement of the colour is made at 625 nm. This will determine sulfide down to about 20 $\mu g\,l^{-1}$.

Sulfide may also be determined by reacting with excess iodine, which oxidizes it to sulfur, and then determining the amount of iodine left, by titration with sodium thiosulfate solution.

Sulfide may also be determined directly with the sulfide ion-selective electrode.[15a, 75]

Sulfides in solution are unstable in that they are very readily oxidized by the ambient air. For this reason to obtain a measurement of soluble sulfide in a sample that contains suspended solids, flocculation with aluminium hydroxide rather than filtration is used to separate them. Water samples in which total sulfides are to be determined are preserved by the addition of zinc acetate (which precipitates the sulfide as zinc sulfide) and filling the bottle completely to exclude air. Precipitation as zinc sulfide is also used to separate the sulfide for a more sensitive determination. Interfering substances in the iodometric determination are reducing agents that also react with iodine, such as sulfite, thiosulfite and some organic compounds. In the methylene blue determination strong reducing agents interfere with the production of the blue colour. Materials which react with iron to form a blue colour, such as ferrocyanide, also interfere.

Hydrogen sulfide in air is determined by precipitating cadmium or zinc sulfide from a solution contained in an impinger or absorber. Cadmium is generally preferred to zinc because its efficiency of collection is greater and the cadmium sulfide is more stable chemically. The precipitated sulfide is determined by one of the methods above. The detection limit is about 1 $\mu g\,H_2S\,m^{-3}$.

Other sulfur acids

It is sometimes necessary to determine the concentration in water samples of other sulfur acids, such as thiosulfate and the polythionates. (A polythionate is $S_nO_6^{2-}$ where n is 3–6.) A common source of these materials is the bacterial action on sulfur-containing effluents from mine and smelter waste drainage. The polythionates can be determined by cyanalysis. When treated with cyanide they will react to form other sulfur acids such as thiosulfate and thiocyanate. The trithionate has been determined by measuring the thiocyanate produced in a colorimetric procedure using the ferric thiocyanate.[76] The total thionates have been measured by reacting the thiosulfate produced in the cyanalysis with excess iodine and determining the residual iodine spectrometrically.[77]

The polythionates can be separated by chromatographic procedures. Wolkoff and Larose[78] used high-speed anion exchange liquid chromatography to separate the thiosulfate, trithionate, tetrathionate and pentathionate ($S_2O_3^{2-}$, $S_3O_6^{2-}$, $S_4O_6^{2-}$ and $S_5O_6^{2-}$) in mining waste water. The mobile phase was 10^{-3} M sodium citrate and the separation was made on the strong anion exchanger Permaphase AAX. The detection was made using Auto Analyzer® equipment. The effluent from the column was segmented, mixed with sodium hydroxide and heated. This converted the polythionates to thiosulfate and sulfite. These were then reacted with cerium(IV) sulfite which was reduced to Ce(III). The fluorescence of the Ce(III) was then measured and gave a chromatogram corresponding to the polythionates and thiosulfate leaving the column. The samples were extracted with chloroform before injection to remove interference from organics. Detection limits of about 0.3 mg l^{-1} were obtained.

A review of the analytical chemistry of the sulfur acids has been made by Szekeres.[79]

Mercaptans—organic sulfur

Mercaptans in air are determined by a reaction similar to that used in the production of methylene blue. The air sample is passed through an absorber or impinger containing mercuric acetate. This solution is treated in the laboratory with ferric chloride and N,N-dimethyl-p-phenylenediamine to produce a colour which is measured at 500 nm (vs 625 nm for sulfides). This method measures the total mercaptans present. It does not distinguish between individual mercaptans, but it appears most sensitive to those of lower molecular weight. The detection limit is about 2 µg m^{-3} mercaptan (expressed as methyl mercaptan).

The organic sulfur compounds such as mercaptans and thioethers can be separated and determined using gas chromatography techniques combined with a flame photometric detector. These techniques are described in Chapter 6.

Total sulfation

The total sulfation in the atmosphere is determined by the use of a lead peroxide 'candle'. Lead peroxide powder and glue are mixed and the paste spread over a solid surface which is exposed to the air. The lead peroxide reacts with the sulfur compounds in the atmosphere, such as hydrogen sulfide, mercaptans, sulfur dioxide, sulfur trioxide, etc., to form lead sulfate. After a certain time the candle is dissolved and the amount of sulfate produced is determined. This may be done by gravimetric precipitation of the barium sulfate or by any other sulfate-measuring technique. A description of the making and measurement of the lead peroxide candle is given in ASTM Standards.[80]

Water samples

No chemical preservatives are added to samples that are to be analysed for the sulfur forms other than sulfide. For sulfide preservation zinc acetate is added (2 ml of 2 N zinc acetate per 100-ml sample) and the bottle is filled completely. All samples for sulfur-form analysis should be stored at low temperature. Samples for the determination of sulfite and other unstable sulfur acids should be immediately cooled and analysed the same day.

Other samples

Biological samples for the determination of sulfur are ashed. Soil and sediment samples are extracted to determine the exchangeable sulfur forms.

Ozone—Oxidants

Ozone in atmospheric samples may be determined by passing the air sample through a solution of potassium iodide. The potassium iodide is oxidized to iodine and this is determined spectrophotometrically by measuring the absorbance of the triiodide ion at 352 nm. Alternatively the iodine can be determined by titration with sodium thiosulfate. The method measures the net oxidizing capacity of the sample: oxidizing agents such as nitrogen dioxide, chlorine and peroxyacyl nitrate (PAN) will give positive interference; reducing agents such as sulfur dioxide and hydrogen sulfide will give negative interference. Of these, nitrogen dioxide and sulfur dioxide constitute the most serious interferences and the method incorporates procedures to correct for them. The determination is probably better described as that of measuring the oxidant content of the atmosphere, where 'oxidant' is defined as: *a material other than nitrogen dioxide which liberates iodine from a buffered potassium iodide solution.* In earlier versions of this method, the absorbing solution of potassium iodide was alkaline. Under these conditions the reduction by sulfur dioxide would not take place until the solution was made acid in the laboratory. The interference from sulfur dioxide was removed by oxidizing it with hydrogen peroxide, then boiling the solution to decompose the excess peroxide. The procedure that is currently recommended[80] is to use a potassium iodide solution buffered at pH 6.8 $\pm$ 0.2. The interfering sulfur dioxide is removed by passing the air through a paper absorber impregnated with chromic oxide. Interference from nitrogen dioxide is best corrected for by simultaneously determining it. The detection limit of this method is 15 µg m^{-3} oxidant as ozone.

The same reaction of oxidation of potassium iodide to iodine is used to determine ozone in water samples.

A method that has been reported as being specific for ozone in the atmosphere is one that uses the reaction with eugenol (4-allyl-2-methoxyphenol).[81] In this method the air being sampled is impinged upon eugenol. The reaction between ozone and eugenol produces formaldehyde and the formaldehyde is absorbed from the air in a second impinger or absorber containing water. This solution of formaldehyde is then analysed by a reverse of the West–Gaeke method for sulfur dioxide (see p. 106).

A method specific for ozone is the cheluminescent reaction with ethylene, which is described in Chapter 8. The other oxidants that are of concern in air sampling are the peroxyacetyl nitrates (PAN) which are formed by the photo-chemical action of sunlight on hydrocarbons and nitrogen oxides. These are determined by gas chromatographic methods.

PARTICULATES

The information required when analysing particulates in air is generally the total mass of the material collected, the particle size and the particle size distribution. The large particles (i.e. those over 10 µm in size) settle very readily and may be collected in a static collector such as a dustfall jar. Particles in the size range between 0.1 and 10 µm are probably the ones of most concern in air pollution work and these are usually collected on a filter. Before use the clean filter is weighed under conditions of humidity and temperature that will be applicable to the loaded filter. The filter is exposed by drawing the air to be sampled through it at a known rate. When the filter has collected the particulates it is re-weighed to determine the amount collected.

Portions of the filter are then taken and examined under a microscope. The pieces of filter are generally immersed in an oil to render the filter transparent and an optical microscope with a magnification up to about 1000 is used to examine the particles. The filter is viewed with a low magnification and the particles larger than a certain size are counted. This is repeated until the highest magnification is reached. The size ranges counted are arbitrary and depend on the types of particulate being viewed. Particles down to about 0.4 µm can be sized by the optical microscope; below this they can be counted but not resolved as to size. The standard statistical methods for determining the area to be viewed to obtain valid size distribution data should be followed. For examination of particles below 0.5 µm an electron microscope must be used, and special filters used to catch these small particles.

If the particulates are collected in an impactor, a size separation at the time of collection can be made by using a cascade impactor. The various fractions collected can be examined on the collection slide as described above.

Depending on the definition of particulates they may or may not include aerosols of liquid materials. If it is desired to measure the mass of aerosols the increase in filter weight will give this, but it will obviously not allow determination of particle size.

In water samples the interest is generally only in determining the mass of suspended matter. The sample is filtered through a medium of a pore size that defines what is suspended and what is soluble. The material caught on this filter is weighed. The generally accepted criterion for particulates in water samples is the inability to pass through a 0.45-µm membrane filter.

Asbestos

Asbestos is a term used to describe a number of hydrated silicates that occur naturally in a fibrous form. The interest in these materials in the environment is one of human health, in that it is known that people who inhale asbestos dust in their work may develop serious diseases such as asbestosis, lung cancer and cancer of the abdominal cavity lining. It has also been demonstrated that people who are only indirectly exposed to asbestos sources may also develop malignancies of the same type. There has been some concern expressed that perhaps exposure to the asbestos levels that are now present in the environment may constitute a health risk and it is desired to determine what are the levels to which people are subjected. In workers who have inhaled asbestos dust, asbestos fibres are found to be widely distributed through their bodies. It is believed that the asbestos fibres can migrate from the lung through the blood and lymphatic system to all parts of the body. Animal studies also show that asbestos that is ingested can pass through the stomach or intestine wall into the blood stream and become lodged in various organs, but it has not been established whether the ingestion of asbestos does in fact constitute a health hazard. The size and shape of the asbestos fibres determine how easily the material can enter cells; the health hazard from asbestos exposure increases the more the material is separated and the fibres shortened. The different chemical compositions of the various asbestos forms may have different effects on the cells they enter. Hence the need in the determination of asbestos in environmental samples, particularly air and water, is to determine not just the mass present but also the size and number of the fibres present and to identify the particular asbestos form.

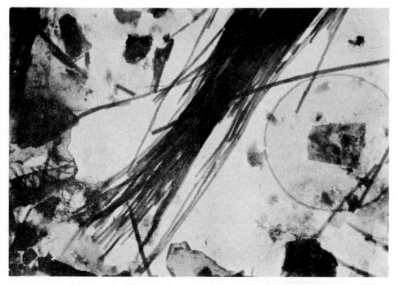

Fig. 4.14. Chrysotile fibres (magnification 30 000).

The fibres may range down to a few hundred nanometers in length and hence the best available technique for their examination is electron microscopy. Identification of the particular asbestos form is made by their morphology, their crystal lattice spacing and their elemental composition. These measurements are made on individual fibres in the electron microscope. A description[82] of the various asbestos forms is as follows: the asbestos minerals can be divided into two main classes on the basis of their crystal structure; serpentine and amphiboles. The sole member of the serpentine class is chrysotile asbestos. This is by far the most common of the asbestos minerals; it accounts for more than 95% of the asbestos fibre produced. There are five asbestos forms that are amphiboles: crocidolite, amosite, anthophyllite, tremolite and actinolite.

The crystal structure of chrysotile asbestos is a layer type. The basis is an infinite silica sheet (Si_2O_5) in which all the silica tetrahedra are pointing in the same direction. Attached to one side of the sheet is a brucite ($Mg(OH)_2$) layer in which two out of three of the hydroxyls are replaced by the apical oxygens of the silica tetrahedra. The result is a double sheet which, because of the mismatch in the dimensions of the silica and brucite sheets, introduces a strain in the structure. To relieve this strain the double sheet is carried with the brucite layer on the outer surface. Hence the chrysotile fibres consist essentially of flat sheets rolled up into a spiral to form a cylinder. Because of this characteristic form, the chrysotile fibres can be identified by their morphology. The appearance of the chrysotile fibres in the electron microscope can be seen in Fig. 4.14.

TABLE 4.1

Compositions and crystal lattice dimensions of asbestos types (from Speil and Lieneweber, Ref. 82)

	Chrysotile	Crocidolite	Amosite	Anthophyllite	Actinolite	Tremolite
SiO_2	38–45	49–53	49–53	56–58	51–56	55–60
MgO	40–43	0–3	1–7	28–34	15–20	21–26
FeO	0–2.6	13–20	34–44	3–12	5–15	0–4
Fe_2O_3	0–1.4	17–20	—	—	0–3	0–0.5
Al_2O_3	0.2–1.5	0–0.2	—	0.5–1.5	1.5–3	0–2.5
CaO	0–1.2	0.3–2.7	—	—	10–12	11–13
K_2O	tr	0–0.4	0–0.4	—	0–0.5	0–0.6
Na_2O	tr	4.0–8.5	tr	—	0.5–1.5	0–1.5
H_2O	12–14	2.5–4.5	2.5–4.5	1.0–6.0	1.5–2.5	0.5–2.5
Unit cell dimension						
a (Å)	5.30	9.75	9.6	18.5	9.85	
b (Å)	9.10	18.0	18.3	18.0	18.1	
c (Å)	7.32	5.3	5.3	5.3	5.3	
γ	93°	103°	105° 50′	—	104° 50′	
Crystal system	monoclinic	monoclinic	monoclinic	orthorhombic	monoclinic	

The basic crystal form of the amphibole minerals is a double silica chain (Si_4O_{11}). All the silica tetrahedra point in the same direction, as in the chrysotile sheet. The chains are paired back-to-back with a layer of hydrated cations in between to satisfy the negative charge of the silica chains. These sandwich ribbons are stacked together in an ordered array. The various minerals in the amphibole group are characterized by the cations occurring in the sandwich; these are principally magnesium, iron, calcium and sodium. The bonding between the ribbons is weak and the crystals are easily cleaved along the outside edges of the sandwiches in a direction parallel to the silica chain direction. This facile cleavage results in the fibrous form of the asbestos minerals.

The range of chemical composition of the different asbestos minerals, and the crystal lattice dimensions are shown in Table 4.1.

For the analysis, the fibres are separated from the sample and ashed. (This procedure is discussed below.) The fibres are then examined in the electron microscope. Both scanning electron microscopes (SEM) and transmission electron microscopes (TEM) have been used in the determination of asbestos. The TEM has a higher resolution than the SEM but requires a more elaborate procedure in order to mount the sample for examination. The particles are viewed and the fibrous material is characterized by its aspect ratio, i.e. the ratio of length to diameter. This should be greater than 3 to 1 in order for the particle to be counted as a fibre. The size distribution and the number of fibres are obtained. The total mass of the fibres is computed from the calculated total volume of the individual fibres and the known density of asbestos. Individual fibres are examined by electron (or X-ray) diffraction. In this technique the electron beam is directed at the fibre and is diffracted by the crystal planes in the fibre. This yields a pattern of spots on the viewing screen; the pattern is characteristic of the crystal form, lattice spacing and orientation of the fibre. From this pattern it is possible to identify the fibre as being a non-asbestos particle, a chrysotile fibre or an amphibole. Unfortunately it is not always possible to obtain an unambiguous diffraction pattern from all the fibres observed. In this case, or if it is desired to classify an amphibole fibre as to asbestos type, an X-ray fluorescence elemental analysis is carried out.

The fibre is irradiated and the resulting X-ray energy spectrum is obtained (see Chapter 2). From this the relative fluorescence peak heights of Na, Mg, Ca, Fe and Si are tabulated. The ratios of the peak heights can be used to supplement the information from the diffraction pattern and to characterize an amphibole. In the work reported by Millette and McFarren,[83] the Na/Mg, Mg/Ca and Fe/Si ratios were used to discriminate between the amphibole types.

Figure 4.15 shows where on a pseudo-ternary composition diagram the different asbestos materials are located. (These positions may be calculated from the data in Table 4.1.)

In any sampling for asbestos it is extremely important that the sample be treated in such a way as to preserve the original asbestos particles in an undisturbed state. Since one of the purposes of the analysis is to determine the particle

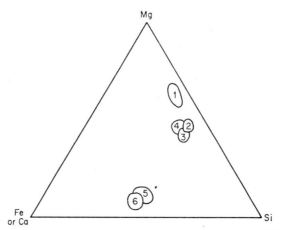

Fig. 4.15. Location of asbestos types on pseudo-ternary composition diagram. 1, Chrysotile; 2, Anthophyllite; 3, Tremolite; 4, Actinolite; 5, Crocidolite; 6, Amosite.

size distribution, the asbestos fibres should not be subjected to any treatment that will degrade them. The procedures that are used for environmental samples are given below.

Air

In determining the levels of chrysotile in air, Rickards[84] took the airborne solids, collected with a Litton high volume sampler, typically from 1000 m³ air, and filtered them through a cellulose nitrate membrane filter of 0.05 μm pore diameter. To improve the filtration rate the samples were coagulated with sodium chloride solution. The separated samples were weighed—a typical sample weighed 100 mg. The filter was then moistened with dibutyl phthalate to avoid loss of fibres by explosive combustion of the filter membrane and ashed at 450 °C for two hours in a glass tube. The resulting ash typically weighed 20 mg. Water was added to the ashed sample and the tube immersed in an ultrasonic water bath for a minimum period of four hours. The sample was then diluted further with water and given ultrasonic treatment for a further eight hours to ensure complete disintegration of agglomerated particles. A drop (10 μl) of this suspension was then placed on a carbon-coated 400-mesh electron microscope grid and examined. The fibres were sized and identified as chrysotile by their morphology and electron diffraction patterns. The weight of chrysotile per cubic metre air was then calculated. It was estimated that the detection limit was 0.1 ng m⁻³. It should be noted that the purpose in this work was to determine the mass present, not the fibre size, and the technique deliberately reduces the asbestos present to small fibrils in order to identify and quantify the chrysotile.

A procedure used by Mueller *et al.*[85] comprised the collection of the particulates by passing the air through a cellulose membrane filter of 0.8 μm pore size. The filter was then ashed in an oxygen plasma. The ash was suspended in a solution of collodion in amyl acetate using an ultravibrator. The collodion film

TABLE 4.2
Results of analysis of air samples (from Pattnaik and Meakin, Ref. 86)

Sample type[a]	Air volume/m^3	Magnification	No. of asbestos fibres counted	Fibre No. m^{-3} × 10^3	Concentration/ g m^{-3} × 10^{-9}	Fibre type
PS	0.135	1 000	97	3 800	200 000	Chrysotile
PS	0.090	3 000	24	1 900	530	Most chrysotile
PS	0.135	3 000	88	5 100	795 000	Chrysotile
PS	0.135	3 000	72	7 800	478 000	Chrysotile
NPS	0.135	1 000	60	2 200	14 300	Most chrysotile, some tremolite
NPS	0.135	1 000	56	400	153 000	Equal chrysotile and tremolite
Demolition site	0.96	1 000	9	11	4 000	Chrysotile and tremolite
Ambient air	10.3	3 000	6 + 2 clumps	2	67	Chrysotile

[a] PS: point source (inside factory). NPS: near point source (20 ft from factory).

was then picked up on an electron microscope grid and on a microscope slide for examination by phase-contrast light microscopy.

In the procedure described by Pattnaik and Meakin,[86] between 0.1 and 15 m^3 of air were drawn through a Millipore® MF-type filter (47 mm diameter, 0.45 μm pore size). The volume of air was determined by the need for avoiding significant superposition of fibres on the filters. The filter was then placed on a glass disc with rubber cement around the periphery. A carbon layer about 100 Å thick was then deposited on the fibre side of the filter. A circular piece of the coated filter was placed carbon side down on a polished beryllium SEM stub in a holder. When TEM examination was planned one or two BeO or SiO substrates were inserted between the Be stub and the filter. The specimen holder was then placed in a vessel and the filter first attacked with acetone vapour for about 30 minutes. The filter swelled slightly and then softened and settled on the Be stub. The filter was then dissolved by immersion in acetone, and after about two hours the acetone was pipetted away. The organic components were then removed by placing the assembly in a low temperature oxygen plasma. This step took about three to four hours. In this way the fibres that were originally in the filter were transferred to a carrier that could be examined in the electron microscope in an essentially unchanged physical form. The results of analysis of airborne asbestos samples are shown in Table 4.2.

Water

The level of asbestos fibres in water is of the order of 10^6 fibres l^{-1}. This level is too low to permit direct examination by electron microscopy, so that a pre-concentration step is necessary. The two methods commonly used are ultracentrifugation and filtration. The technique used by Cunningham and Pontefract[87] in the examination of beverages and drinking water was to centrifuge about a 400-ml sample. The fibres obtained were dried at 100 °C and then ashed at 500 °C. The ash was taken up in distilled water and the fibres re-suspended with an ultravibrator.

This technique has been questioned since it has been shown that there is a loss of fibres by the combustion of the filter in the high temperature ashing. This difficulty is overcome using the low temperature plasma oxidation.

Separation by filtration is the alternative procedure, and was used in the methods described by Beaman and File.[88] The water samples were filtered through an 0.22-μm cellulose ester filter. The filter was dried and a circular section cut from it. This section was placed on a TEM grid and the filter dissolved with acetone vapour in a condensation washer. It was found that some of the fibres were lost in this dissolving procedure; the average loss was 50% for amphibole fibres and 10% for chrysotile fibres. Corrections were made for these losses in the calculation of the final concentrations.

After the asbestos fibres have been separated by filtration the filter can be placed in water and the fibres resuspended by sonification. A drop of the concentrated suspension can then be placed on the electron microscope grid and the water evaporated. This avoids the loss of fibres experienced when filters are dissolved (above) but the uneven drying of the drop tends to concentrate the fibres around its outside edges. In a technique described by Mudroch and Kramer[89] the fibres in a water sample are centrifuged directly on to an electron microscope grid placed in the bottom of the centrifuge tube.

Some water samples are sufficiently 'clean' to allow direct examination of the separated fibres. However, in many cases there is a sufficient number of non-asbestos particles, particularly organics, that are separated with the asbestos to make a direct measurement very difficult. In this case it is necessary to remove the organics by ashing. The ashing procedure introduces the problem that after the ashing process the fibres tend to be clumped in the remnants of the ash, making examination difficult. This clumping can be diminished by treating the suspension in water by ultrasonification, but any such process may introduce artifacts into the analysis.

Solid samples

Solid samples can be treated according to the method recommended by UICC (International Union Against Cancer).[90] The material is suspended in a solution of Gurr's collodion dissolved in an ether–alcohol mixture. A drop of this suspension is allowed to fall on the surface of the distilled water. The resulting film may then be lifted off on a cover slip for examination with a light microscope or used to prepare an electron microscope grid.

REFERENCES

1. L. Snyder, J. Levine, R. Stoy and A. Conetta, *Anal. Chem.* **48**, 942A (1976).
2. *Standard Methods for the Examination of Water and Wastewater*, 14th Edn. APHA-AWWA-WPCF (1975).
3. L. J. Lionnel, *Analyst* **95**, 194 (1970).
4. P. D. Goulden and Y. P. Kakar, *Water Res.* **10**, 491 (1976).
5. E. Goldman, S. Taormina and M. Castillo, *J. Am. Water Works Assoc.* **67**, 14 (1975).
6. P. Hulthe, L. Uppstrom and G. Ostling, *Anal. Chim. Acta* **51**, 31 (1970).
7. F. H. Rainwater, *J. Am. Water Works Assoc.* **51**, 1046 (1959).
8. A. A. Levinson, *Water Res.* **5**, 41 (1971).
9. J. Bassett and P. J. Matthews, *Analyst* **99**, 1 (1974).
10. H. James and G. H. King, *Automation in Analytical Chemistry*, Mediad Inc., New York, 1967, p. 123.
11. B. K. Afghan, P. D. Goulden and J. F. Ryan, *Water Res.* **6**, 1475 (1972).
12. D. M. Zall, D. Fisher and M. D. Garner, *Anal. Chem.* **28**, 1665 (1956).
13. R. J. Bertolacini and J. E. Barney, *Anal. Chem.* **29**, 281 (1957).
14. B. K. Afghan, R. Leung, A. V. Kulkarni and J. F. Ryan, *Anal. Chem.* **47**, 556 (1975).
15. J. Koryta, *Anal. Chim. Acta* **61**, 329 (1972).
15a. P. L. Bailey, *Analysis with Ion-selective Electrodes*, Heyden, London, 1976.
16. I. Sekerka, J. F. Lechner and R. Wales, *Water Res.* **9**, 663 (1975).

17. W. Likussar, H. Raber, H. Huber and D. Grill, *Anal. Chim. Acta* **87**, 247 (1976).
18. M. J. Fishman and M. W. Skougstad, *Anal. Chem.* **35**, 146 (1963).
19. R. C. Harriss and H. H. Williams, *J. Appl. Meteorol.* **8**, 229 (1969).
20. E. J. Newman, *Analyst* **96**, 384 (1971).
21. S. S. Yamamura, M. A. Wade and J. H. Sikes, *Anal. Chem.* **34**, 1308 (1962).
22. Industrial Method No. 129-71W, Technicon Instrument Corporation, Tarrytown, N.Y.
23. J. E. Harwood, *Water Res.* **3**, 273 (1969).
24. I. Sekerka and J. F. Lechner, *Talanta* **20**, 1167 (1973).
25. J. A. Cooke, M. S. Johnson and A. W. Davison, *Environ. Pollut.* **11**, 257 (1976).
26. P. Casapieri, R. Scott and E. A. Simpson, *Anal. Chim. Acta* **49**, 188 (1970).
27. P. D. Goulden, B. K. Afghan and P. Brooksbank, *Anal. Chem.* **44**, 1845 (1972).
28. A. Conetta, J. Jansen and J. Saltpeter, *Pollut. Eng.,* Jan. 1975, p. 36.
29. N. P. Kelada, D. T. Lordiand and C. Lue-Hing, *Cyanide Species Methodology in Water, Wastewater and Sediments*, Technicon International Congress, New York, December, 1976.
30. I. Sekerka and J. F. Lechner, *Water Res.* **10**, 479 (1976).
31. H. A. C. Montgomery, D. K. Gardiner and J. G. G. Gregory, *Analyst* **94**, 284 (1969).
32. M. B. Jacobs and S. Hochheiser, *Anal. Chem.* **30**, 426 (1958).
33. J. Mulik, R. Fuerst, M. Guyer, J. Meeker and E. Sawicki, *Int. J. Environ. Anal. Chem.* **3**, 333 (1974).
34. Instruction Manual IM46, Orion Research, Cambridge, Mass.
35. P. G. Brewer and J. P. Riley, *Deep-Sea Res.* **12**, 765 (1965).
36. Industrial Method No. 100-70W, Technicon Instrument Corporation, Tarrytown, N.Y.
37. P. D. Goulden and Y. P. Kakar, *Anal. Lett.* **8**, 763 (1975).
38. L. J. Forney and J. F. McCoy, *Analyst* **100**, 157 (1975).
39. A. Fleck, C.R.C. Crit. Rev. in *Anal. Chem.* 141 (1974).
40. R. Sawyer and L. M. Grisley, *Automation in Analytical Chemistry*, Mediad, New York, Vol. I, 1967, p. 347.
41. W. H. Evans and B. F. Partridge, *Analyst* **99**, 367 (1974).
42. W. M. Crooke and W. E. Simpson, *J. Sci. Fd Agric.* **22**, 9 (1971).
43. O. Elkei, *Anal. Chim. Acta* **86**, 63 (1976).
44. A. Henriksen, *Analyst* **95**, 601 (1970).
45. B. K. Afghan, P. D. Goulden and J. F. Ryan, *Advances in Automated Analysis*, Vol. II, Thurman Assoc., Miami, Florida, 1970, p. 291.
46. S. R. Carlberg (Ed), *Cooperative Research Report*, Series A, No. 29, p. 73, International Council for the Exploration of the Sea, Charlottenlund, Denmark (1972).
47. D. K. Albert, R. L. Stoffer, I. J. Oita and R. H. Wise, *Anal. Chem.* **41**, 1501 (1969).
48. R. E. Parks, *Analysis for Chemically Bound Nitrogen using Pyrochemiluminescence*, Paper No. 184a, Pittsburgh Conference on Analytical Chemistry and Applied Spectroscopy, March 1976.
49. A. Henriksen, *Analyst*, **90**, 29 (1965).
50. D. L. Johnson, *Environ. Sci. Technol.* **5**, 411 (1971).
51. P. D. Goulden and P. Brooksbank, *Limnol. Oceanogr.* **19**, 705 (1974).
52. A. G. Fogg and K. S. Yoo, *Anal. Lett.* **9**, 1035 (1976).
53. F. H. Rigler, *Limnol. Oceanogr.* **13**, 7 (1968).
54. ASTM Standards, Part 31, American Society for Testing and Materials, Philadelphia, Pa. (1976).
55. L. P. Tyler and A. F. Biles, *Advances in Automated Analysis*, Vol. II, Thurman Assoc., Miami, Fla., 1970, p. 313.

56. W. F. Milbury, V. T. Stack and F. L. Doll, *Advances in Automated Analysis*, Vol. II, Thurman Assoc., Miami, Fla., 1970, p. 299.
57. F. A. J. Armstrong and S. Tibbitts, *J. Mar. Biol. Assoc. U.K.* **48**, 143 (1968).
58. K. Grasshoff, *Automation in Analytical Chemistry*, Vol. I, Mediad Inc., N.Y., 1967, p. 573.
59. P. D. Goulden and P. Brooksbank, *Anal. Chim. Acta* **80**, 183 (1975).
60. S. J. MacKay, *Int. J. Environ. Anal. Chem.* **4**, 33 (1975).
61. R. A. Scott and G. P. Haight, *Anal. Chem.* **47**, 2439 (1975).
62. N. Yoza, K. Kouchiyama, T. Miyajima and S. Ohashi, *Anal. Lett.* **8**, 641 (1975).
63. G. W. Heinke and H. Behmann, *Advances in Automated Analysis*, Vol. II, Mediad Inc., New York, 1969, p. 115.
64. G. W. Fuhs, *Int. J. Environ. Anal. Chem.* **1**, 123 (1971).
65. S. C. Chang and M. L. Jackson, *Soil Sci.* **84**, 133 (1957).
66. J. D. H. Williams, J. M. Jaquet and R. L. Thomas, *J. Fish. Res. Board Can.* **33**, 413 (1976).
67. J. D. H. Williams and T. W. Walker, *Soil Sci.* **107**, 213 (1969).
68. N. C. Mehta, J. O. Legg, C. A. I. Goring and C. A. Black, *Soil Sci. Soc. Amer. Proc.* **18**, 443 (1954).
69. K. I. Aspila, H. Agemian and A. S. Y. Chau, *Analyst* **101**, 187 (1976).
70. J. K. Syers, R. F. Harris and D. E. Armstrong, *J. Environ. Qual.* **2**, 1 (1973).
71. Industrial Method No. 118-71W, Technicon Instrument Corporation, Tarrytown, N.Y.
72. R. M. Carlson, R. A. Rosell and W. Vallejos, *Anal. Chem.* **39**, 688 (1967).
73. J. Mulik, R. Puckett, D. Williams and E. Sawicki, *Anal. Lett.* **9**, 653 (1976).
74. F. P. Scaringelli, B. E. Saltzmann and S. A. Frey, *Anal. Chem.* **39**, 1709 (1967).
75. E. W. Baumann, *Anal. Chem.* **46**, 1345 (1974).
76. D. P. Kelly, L. A. Chambers and P. A. Trudinger, *Anal. Chem.* **41**, 901 (1969).
77. T. Koh and K. Taniguchi, *Anal. Chem.* **45**, 2018 (1973).
78. A. W. Wolkoff and R. H. Larose, *Anal. Chem.* **47**, 1003 (1975).
79. L. Szekeres, *Talanta* **21**, 1 (1974).
80. ASTM Standards, Part 26, American Society for Testing and Materials, Philadelphia, Pa. (1976).
81. S. L. Sachdev, J. P. Lodge and P. W. West, *Anal. Chim. Acta* **58**, 141 (1972).
82. S. Speil and J. P. Lieneweber, *Environ. Res.* **2**, 166 (1969).
83. J. R. Millette and E. F. McFarren, *Scanning Electron Microsc. Proc.* **1**, 451 (1976).
84. A. L. Rickards, *Anal. Chem.* **45**, 809 (1973).
85. P. K. Mueller, G. R. Smith, L. M. Carpenter and R. L. Stanley, *Proc. Electron Microscopy Soc. Amer.* **30**, 356 (1972).
86. A. Pattnaik and J. D. Meakin, *Scanning Electron Microscopy Proc.* **1**, 441 (1976).
87. H. M. Cunningham and R. Pontefract, *Nature (London)* **232**, 332 (1971).
88. D. R. Beaman and D. M. File, *Anal. Chem.* **48**, 101 (1976).
89. O. Mudroch and J. R. Kramer, *Proc. Electron Microscopy Soc. Amer.* 526 (1974).
90. M. I. Skikne, J. H. Talbot and R. E. G. Rendall., *Environ. Res.* **4**, 141 (1971).

CHAPTER 5

RADIONUCLIDES

TYPES OF RADIOACTIVITY

Radioactivity is a characteristic of elements that have unstable nuclei. In going to a more stable state the nucleus releases energy, either as a particle with kinetic energy or as electromagnetic radiation. The progress to a stable state may be made in one step or it may require several sequential processes of energy release. The common types of radioactive emissions are: alpha (a), beta (β^-), positron (β^+), gamma (γ), electron capture (e.c.) or neutron (n).

Alpha radiation is the emission of doubly charged helium nuclei (He^{2+}). The energy is in the form of the kinetic energy of the particles.

Beta radiation is the emission of electrons moving at high velocity; the energy released is in the kinetic energy of the particles. Positron emission is similar to β radiation except that the particles are positively charged.

Gamma radiation is electromagnetic radiation and its energy is quantized as photons with a characteristic frequency.

Electron capture is the situation in which the nucleus becomes more stable by capturing an electron from one of the inner shells around the atom. This produces an instability in the electron shells and their rearrangement leads to the emission of X-rays.

Neutron emission is the loss of one unit of mass in an uncharged state; the energy is represented by the kinetic energy of the neutron.

These various radiations produced in radioactive decay lose their energy to the surrounding matter in characteristic ways. They are, however, all high energy radiation and, as are X-rays, they are ionizing radiations, i.e. as they interact with the atoms of the surrounding materials they displace electrons from them, the electrons being displaced with sufficient kinetic energy to themselves cause further ionization. The a particles, being large charged particles moving comparatively slowly, react very strongly with the electrons of the atoms in the surrounding materials. They lose their kinetic energy over a short path causing intense ionization for a small distance. The β particles are also charged, but their mass is less than that of α particles and their velocity is higher. Hence they tend to penetrate matter more deeply than α particles giving a less dense ionization. Gamma radiation has a much smaller probability of interacting with the electrons in the surrounding material than the particulate radiations, and hence penetrates much farther before losing its energy. The energy of the γ radiation may be dissipated in one of three ways. These are:

(a) The photoelectric effect, where the energy of the γ ray is transferred to a single atom. This results in the atom emitting a single photoelectron. Any excess energy of the γ ray over that required to release the electron is imparted to the electron as kinetic energy. This electron then loses this kinetic energy in causing ionization in the surrounding material.
(b) Compton scattering, where the γ ray interacts with the electrons of the absorbing material to lose a fraction of its energy in each interaction.
(c) Pair production, where the energy of the γ ray is converted to matter, a positron and an electron being produced.

Neutrons do not react with the electrons in the surrounding material but can penetrate to the nuclei of the atoms. There they can interact, either with a loss of energy through scattering of the neutron, or with absorption of the neutron. If the neutron is absorbed, a new unstable nucleus is formed which later decays.

The characteristic properties of the radiations produced in the disintegrations are used in the identification and measurement of radioactive materials.

The α particles which are produced by a particular nuclide all have essentially the same energy, characteristic of that nuclide. This energy is in the range 4–8 MeV. The γ radiation emitted has energy levels that are characteristic of the particular nuclide and decay process. These are usually in the range 0–2.5 MeV. The β particles emitted are not at a single energy level; their energies range from zero to a maximum that is characteristic of the nuclide.

MEASUREMENT OF RADIATION

Radiation is measured by observing the interactions which occur when it is absorbed by a material. These interactions may produce ionization, light, electricity or heat. The three common types of detectors which are used are: ionization detectors, scintillation detectors and semiconductor detectors.

Ionization detector

If the radiation is absorbed in a gas in a chamber, ionization of the gas occurs and the positive and negative ions can be collected by charged electrodes. The current that flows between the electrodes is a measure of the ionization produced in the gas. This, in turn, is a function of the chamber size, the number of disintegrations occurring per second, the energy released in each disintegration, the fraction of this energy that is absorbed in the chamber, and the ionization energy of the gas.

For high efficiency it is better that the radioactive material be inside the chamber so that none of the energy of the radiation is lost in penetrating the chamber wall. This is particularly true when measuring a particles since their penetrating power is low. For convenience in not having to de-mount the chamber, some are used with a thin window at one end. The radiation should also be completely absorbed in the chamber to obtain high efficiency. In the case of γ radiation, which is difficult to absorb, the gas in the chamber may be at increased pressure to increase its density. The type of gas used in the chamber is important. When air is used the oxygen in it tends to combine with the electrons produced during the ionization to form negative oxygen ions. These heavy ions are more difficult to collect than free electrons and so the electrical output obtained is reduced. A better output is obtained when an inert or noble gas is used.

When ionization occurs in the chamber no current flows until a certain potential is applied. From this point the current increases as the potential increases until all the ionized material is being collected. Above this potential (the saturation potential), there is no increase in current with potential, and in this range of operation small changes in applied potential do not affect the current being measured. When measuring low energy radiation the ionization current becomes difficult to measure accurately and instead of measuring the current the number of ionizing events are counted. Each ionizing event produces a number of electrons which are collected and produce a small pulse of current. The number of these pulses are counted. In this pulse mode of operation the filling gas is important, in that the ions produced must be collected quickly to give a sharp pulse. To improve the sensitivity of this pulse measurement the potential across the collection electrodes in the chamber is increased to the point where the electrons being collected are highly accelerated. These electrons then travel at sufficiently high velocity to cause further ionization in the filling gas. Hence each primary ionization results in a chain of secondary ionizations. An amplification of up to 10^5 can be obtained in the signal obtained. The counters operating in this way are called proportional counters since the size of the output pulse obtained is proportional to the original ionization energy. The size of the output pulse can then be used to measure the energy of the radiation, in addition to measuring the total number of counts. However, the amplification obtained is a function of the potential applied and this must be controlled carefully if energy discrimination is to be made.

For the discrimination of the energy of a particles a gridded ion chamber is used. This consists of an ion chamber fitted with a collimator between the source and the collection plate, and a grid to shield the collector from the effects of positive ions.

When the potential is increased above that used in proportional counting, a region is reached where the potential gradient around the anode is so high that the discharge of secondary ionization spreads along the electrode. Then the size of the output pulse is independent of the original ionization energy; each ionizing event that occurs results in a complete discharge at the anode, and hence a large output. This is known as the Geiger region and is used in the Geiger counter detectors. One problem with the use of the high collection voltage is that, at the cathode, the positive ions are also accelerated and they will cause the formation of photoelectrons as they strike the cathode. These photoelectrons will be collected in the same way as the original 'ionization' electron so that spurious pulses will be obtained after each ionization event. This difficulty is overcome by adding a quenching gas which absorbs energy from the positive ions so that they will no longer cause ionization. The gas commonly used in the Geiger chambers is a halogen such as bromine. This is disrupted by the positive ions to form bromine atoms, which then recombine to reform the original bromine molecules.

Scintillation detectors

In a scintillation detector the radiation reacts with the absorbing material to displace electrons. On the subsequent rearrangement of the electrons the energy is released as a light photon. The early techniques used a screen coated with zinc sulfide and the light flashes were observed visually. A number of other materials such as the organics, anthracene or stilbene, and inorganic crystals such as sodium iodide are now used as scintillation phosphors, but zinc sulfide is still the most widely used phosphor for a counting. The light photons emitted by the phosphor are measured with a photomultiplier. In order that the phototube may see the light emitted, the phosphor must be transparent. In the case of a particle counting the phosphor can be spread as a thin layer on, for example, Mylar® film, since the a particles are absorbed very readily. In the case of β and γ radiation, there must be provided a sufficient mass to absorb a significant fraction of the radiation. Single crystals are used, such as a large (e.g. 3 in. diameter, 3 in. thick) cylinder of sodium iodide. Other systems use the phosphor dissolved in a plastic or, as in liquid scintillation, the phosphor may itself be suspended in a solution containing the nuclide. In the case of inorganic phosphors the pure substances do not work very well. Silver is added to zinc sulfide to activate it. A small quantity of thallium is added to the sodium iodide phosphors. The sodium iodide crystal is then described as a thallium-activated sodium iodide crystal (NaI(Tl)).

As in the case of the proportional counter, the magnitude of the light emitted when radiation strikes the phosphor is dependent on the energy of the incident radiation. This is employed in the determination of the γ ray spectrum of nuclides using a sodium iodide crystal. The photomultiplier is connected to a multichannel analyser which 'sorts' the pulses seen into their respective energy channels.

Semiconductor detectors

The semiconductor detector operates like an ionization chamber, except that, instead of absorbing the radiation in a gas and collecting the ions formed, the radiation is absorbed in a solid semiconductor. When the radiation interacts with the crystal of the semiconductor, it produces electron–hole pairs analogous to the electron–positive ion pair formed in the ionization chamber. The holes behave like positive electrons and by the application of a potential across the semiconductor, the two charges can be collected for measurement. The big advantage of the semiconductor detector is that the energy required to produce an electron–hole pair is much less than that required to produce an ion pair in a gas chamber or a photon in the scintillation detector. This results in a much better resolution being obtained when using the semiconductor detector for the determination of a radiation spectrum. The drawback to its use is that it is limited in size compared to a single crystal of sodium iodide.

Detectors made of silicon are used for α counting and α spectrometry. They are also used for β spectrometry. For γ spectrometry a germanium semiconductor is used, since germanium, with a higher atomic number, is a more efficient absorber of γ photons. The semiconductors with the highest performance are those in which lithium is introduced into the crystal. It does not replace any of the atoms in the lattice but positions itself between the lattice atoms. This alters the electrical characteristics of the semiconductor so that a thick crystal can be made which still has good collection properties. The lithium is said to be 'drifted' into the crystal and the semiconductor is described as a lithium-drifted germanium semiconductor, Ge(Li). Because the lithium tends to escape from the crystal at room temperature the Ge(Li) semiconductor is kept at liquid nitrogen temperature ($-200\,°C$).

DETERMINATION OF RADIOACTIVE MATERIAL

In the determination of environmental radioactivity levels the concern is primarily one of human health. It is known that ionizing radiation, by its disruptive effect, will cause changes to take place in the functioning of cells within the body. This effect is dependent on the amount of radiation received down to very low levels; there does not appear to be a threshold level below which there is no effect. Radiation received from outside the body can be monitored by a film badge or other personal radiation-measuring device. This is a routine safety procedure in industrial operations involving exposure to radiation. The amount of whole-body radiation a person may receive is limited by health regulations. The hazard from radioactive materials in the environment, particularly in water, food and air, is that they are ingested and metabolized and become incorporated within the body. The radiation produced then irradiates the body from the inside. In the case of the radiation of low penetrating power, such as α particles and soft β particles, the energy of the ionizing radiation is dissipated in a very small volume of body cells thus giving a very concentrated dose. Perhaps the most severe health risks are those nuclides such as plutonium and strontium-90 which become incorporated into the bone and there irradiate the bone marrow. Hence each nuclide presents a specific hazard based on its chemistry, its mode of decay and its specific activity. While a measure of the gross radioactivity in a sample is useful as an indication of a possible problem, the identification of the particular nuclides present is usually required in a determination of radioactive materials in an environmental sample.

The unit of measurement used is a curie, which is that amount of material undergoing 3.7×10^{10} disintegrations per second. The normal concentrations that are of concern in, for example, drinking water for such nuclides as ^{90}Sr are in the picocuries per litre range. One picocurie (pCi) is the amount of material undergoing 2.22 disintegrations per minute. This means that the amount of ^{90}Sr that is of concern is of the order of 10^{-14} g l^{-1}.

One way to determine the radioactive materials in a sample is to concentrate the sample, look at the spectrum of radiation coming from the whole sample, and from this spectrum identify the nuclides present. In the case of the nuclides which are γ emitters this is done using γ spectrometry. The other way to determine specific nuclides is by chemical separation. They are separated from the matrix and other nuclides by their chemical properties, and their concentration is determined by measuring the radiation from the separated (and concentrated) material. These concentrates can also be examined by γ spectrometry to identify particular nuclides.

SAMPLE COLLECTION AND PREPARATION

Sample collection for radioactivity determinations is no different from that used for any other determination, particularly in the case of water and solid samples. In the case of air samples, if the particulates are to be determined by direct counting or γ spectrometry, then it is convenient to use a filter that will fit directly into the detection device. If the particulates are to be a-counted a filter that collects the particles on its surface should be used so that there is less absorption when the filter is counted. This is not as important when measuring β and γ contamination. The collection for tritium contamination in the air is usually by collecting the water vapour from the air either by a cold trap or by an adsorbent.

Sample preparation for chemical separations is the same as for other types of analysis. Organic samples are ashed to remove organic matter; the same choice of ashing methods is made as in preparation for, for example, atomic absorption measurements. Soil and sediment samples are extracted or dissolved, depending on the particular nuclide to be determined. Air particulates are treated the same way as soil samples. One difference in the sample preparation from that used for other analyses is the necessity of incorporating the carrier in the sample (see below). Thus after ashing an organic sample it may be preferred to carry out a fusion, rather than an extraction, on the ash to ensure that the carrier is in equilibrium with the sample. Similarly a soil or sediment may be fusion-treated rather than extracted for the same reason.

γ-SPECTROMETRY

The energy spectrum of the γ radiation from a mixture of radionuclides can be obtained using either a NaI(Tl) crystal and a photomultiplier or a Ge(Li) detection system. The NaI(Tl) crystal because of its larger size has a higher geometric efficiency in collecting the γ radiation. However, as described above, because of the lower energy requirement of the Ge(Li) semiconductor for the production of an electron–hole pair, the Ge(Li) detector has better resolution. Hence the usual choice for examining environmental samples is to use the Ge(Li) system. The levels of the radioactive material found are generally quite low and a preconcentration step is necessary for such samples as water. These are generally concentrated by evaporating a volume of several litres. One technique evaporates the water on a plastic sheet in a dish under a heat lamp. When the water is evaporated the plastic sheet containing the solid material is rolled up and used as the sample in the spectrometer. Air samples may be collected on a carbon filter which may be directly examined. Biological samples are ashed to reduce their volume.

In a typical 'large-scale' operation the detector system is housed in a well shielded room, since the limit of detection is governed by the amount of background radiation received. In the description of low-background Ge(Li) detection systems for radio-environmental studies by Camp *et al.*[1] the limits of such systems are discussed. In addition to the radiation received from the surrounding area, they measured the natural radioactivity of materials of construction. They showed that aluminium and silicon contain appreciable amounts of radioactive material and so the metal housing in their system was made of magnesium. Pyrex® glass was also shown to contain some ^{40}K. The other limit on the detection was the obscuring of the peaks of the lower-energy γ rays by the Compton continuum from higher-energy elements. To obtain high sensitivity a Ge(Li) detector housed in magnesium was operated in anticoincidence with a large plastic detector. The whole system was placed in a lead and steel armour plate shield located in a counting room below ground level. It was shown that, using a 1000-minute count, the system would measure 20 fCi g^{-1} of ^{130}Cs in a soil sample of 250 g with a precision of 15%.

Most operating systems do not have this high sensitivity, but will measure environmental levels of radionuclides in concentrated samples. The systems work best for nuclides which are multiple γ emitters, since they can be more positively identified. For the nuclides which emit only one γ energy, accurate calibration of the energy measurement has to be made and, if possible, some assessment of the halflife.

CHEMICAL SEPARATION

In carrying out chemical separations the amount of material that is being determined is generally so small that it is not feasible to carry out directly chemical operations that depend on mass, e.g. precipitation reactions. The solubility products necessary for precipitation are many orders of magnitude larger than the concentration being determined. Some separations that are independent of mass, but depend on concentration equilibriums such as solvent extraction and ion exchange, may be able to be used directly. The technique used to enable the 'mass' reactions to operate in the separation is to add a known amount of an inert isotope of the element being determined, usually in the milligram range. This added amount is called a carrier and the assumption made in its use is that it will behave chemically exactly like the nuclide being determined. In order to do this it must be equilibrated with the other nuclide in the same chemical form. Thus at the point where it is added, some rather drastic chemistry is carried out to ensure that this equilibration does take place. Some elements do not have inert isotopes (radium, uranium and thorium, for example). In this case an element of similar chemical properties can be used—lead or barium will 'carry' radium as the sulfate. One carrier will sometimes also act for a group of elements —iron will act as a carrier for the rare earth hydroxides. Because of the small amount of nuclide present, it may carry along with other materials that are precipitated, although chemistry would predict that it should be soluble. This is generally an adsorption phenomenon and may also occur on the walls of the containers used for carrying out the reaction. For this reason it is often an advantage to add carrier even if the separation being carried out is apparently independent of mass. The other advantage in using a carrier is that the amount recovered at the end of the separation can be measured, and hence the efficiency of the separation process determined. The recovery of the carrier is often determined gravimetrically and, if it is, the form of the material recovered must be suitable for gravimetric measurement. Gravimetric determinations are the least desirable, however, because they are non-specific for the element. Better methods for determining recoveries are those which can be element-specific such as volumetric or colorimetric measurements. In some cases it is possible to add a carrier that can be determined by a radiation measurement; for example in the determination of strontium-90, strontium-85 can be added and its recovery determined by γ counting.

The detailed procedures for determining some of the environmentally important nuclides are as follows.

Tritium

Tritium is a β emitter, with a maximum β energy of 0.018 MeV. It has a half-life of 12.26 years. It is determined by liquid scintillation counting. The sample is distilled, both to remove organic materials which may have a quenching action in the counting and also to separate it from other radioactivity.

In the procedure[2] the sample (about 30 ml) is distilled from potassium permanganate to near dryness. The distillate is collected and 4 ml is mixed with 16 ml of scintillation solution in a scintillation vial. The vial is kept in the dark for several hours to dark-adapt it and then it is counted in a scintillation spectro-meter. The efficiency of the counter is determined by counting standards. The scintillation solution is made by dissolving 4 g of PPO (2,5-diphenyloxazate), 0.05 g of POPOP (1,4-di-2-(5-phenyloxazolyl)benzene) and 120 g of naphthalene in 1 litre of 1,4-dioxane.

If concentration of the sample is desired it may be concentrated by a factor of about 10 by electrolysing the sample under either acid or alkaline conditions. The tritium is concentrated in the liquid left in the electrolysis cell, although not quantitatively. The enrichment factor obtained is determined by electrolysing standards.

Strontium

Strontium-89 and strontium-90 are the two isotopes of concern in environmental samples. They are both β emitters; neither of them emits γ radiation. Strontium-89 has a β energy of 1.46 MeV and a halflife of 52 days. Strontium-90 has a β energy of 0.54 MeV and a halflife of 28.1 years. Yttrium-90, which is the daughter product of ^{90}Sr and which is the nuclide actually determined when measuring ^{90}Sr, is a β emitter with a β energy of 2.27 MeV and a halflife of 64 hours. It also emits γ radiation.

The strontium isotopes are separated from other nuclides by a chemical separation. A strontium carrier, a barium carrier, the alkaline earths and the rare earths are precipitated as the carbonates. (The barium carrier is added since there is potential interference from radioactive barium.) The precipitate is dissolved and strontium nitrate and barium nitrate precipitated from fuming nitric acid. This precipitate is redissolved, the rare earths are precipitated as the hydroxides and the barium precipitated as the chromate. This precipitation also removes any interfering radium and lead. The strontium fraction then contains the ^{89}Sr and ^{90}Sr and the total radiostrontium is determined by counting this. It is usually desired to determine the ^{90}Sr, as this constitutes the most severe health hazard because of its long halflife. Since the isotopes cannot be separated chemically and they are not γ emitters, use is made of the fact that ^{89}Sr decays to yield ^{89}Y which is a β emitter. After a delay of at least two weeks to allow the growth of ^{90}Y the yttrium is separated from the strontium either by extraction or by precipitation, and counted. This gives a measure of the ^{90}Sr in the separated strontium fraction.

The procedure followed for analysing water is as follows:[2] to 1 litre of filtered water is added 2 ml of concentrated nitric acid and a barium and strontium carrier (20 mg each of Ba and Sr). The solution is heated to boiling and sodium hydroxide and sodium carbonate are added. The precipitate is allowed to settle and is then separated by decantation and centrifuging. This precipitate is dissolved in 4 ml of concentrated nitric acid and then 20 ml of fuming nitric acid is added and the tube is cooled and centrifuged. The precipitated barium and strontium nitrates are redissolved and the fuming nitric acid precipitation is repeated. It may be repeated again if the original sample contains a large amount of calcium. The precipitate is then washed with acetone and redissolved in water. A rare earth carrier (Ce, Zr, Fe) is added and the rare earth hydroxides precipitated with ammonia. The solution is centrifuged, the supernatant collected and the rare earth precipitation repeated. From the time of the rare earth precipitation the ingrowth period of ^{90}Y begins, hence the procedure should not be delayed unduly after this point. The supernatant liquid is buffered with sodium acetate buffer to pH 5.5 and sodium chromate solution is added to precipitate barium chromate. The strontium is precipitated as carbonate; the precipitate is washed and separated, either by filtration or by drying under a heat lamp on a tared pan. The amount of precipitate is weighed to determine the recovery of the strontium carrier and β counted to determine the total strontium level (^{89}Sr + ^{90}Sr). If ^{85}Sr was used as the carrier the recovery can be determined by γ counting.

The β counting may be carried out using an internal proportional counter or a counter (proportional or Geiger–Muller) fitted with a thin end window. To avoid loss of the precipitate and possible contamination of the counter when counting the precipitate, it may be covered with a thin Mylar® film using the plastic ring arrangement available for this purpose.

To determine ^{90}Sr the precipitate of strontium carbonate obtained above is stored for at least two weeks to allow the ingrowth of ^{90}Y. The yttrium is then separated by solvent extraction or by precipitation as the oxalate.

In the solvent extraction procedure the strontium precipitate is dissolved in nitric acid and an yttrium carrier added. The strongly acid solution is then extracted with two 5-ml portions of tributyl phosphate. The organic layer is washed with nitric acid and the yttrium is back-extracted into dilute acid. This solution is evaporated for counting of the ^{90}Y. In the precipitation procedure the strontium carbonate is dissolved in acid, the yttrium carrier is added and the yttrium precipitated as the hydroxide. This hydroxide is separated and redissolved, the yttrium then being precipitated as the oxalate, which is mounted and counted. If desired the supernatant from the yttrium hydroxide precipitation may be recovered and stored for a further period for a second ingrowth of ^{90}Y.

Strontium is not volatile and dry-ashing of biological samples can be carried out. Solid and sediment samples, and ash from biological samples, are fused with sodium carbonate. The melt is leached with water and any precipitate left is dissolved in nitric acid. The alkaline earths are precipitated with phosphoric acid. The precipitate is then dissolved in nitric acid and treated as for a water sample above. For wet-ashing of biological samples nitric acid is commonly used. Soil and sediment samples may also be leached sequentially with sodium hydroxide and hydrochloric acid.

Radium

There are four naturally occurring radium isotopes: ^{223}Ra, ^{224}Ra, ^{226}Ra and ^{228}Ra. Radium-223 is a member of the actinium series whose parent is ^{235}U; radium-224 and ^{228}Ra are members of the thorium series whose parent is ^{232}Th; ^{226}Ra is a member of the uranium series whose parent is ^{238}U. Radium-228 is a β-emitter which eventually decays to ^{224}Ra by way of one of its daughters (^{228}Th), which is an α-emitter. However, the contribution of ^{228}Ra to the α activity is negligible because of the 1.9 year halflife of ^{228}Th. The other three isotopes are α-emitters and they produce daughters which are also α-emitters. The decay schemes for the radium isotopes are rather complex and are shown in Fig. 5.1.

Actinium Series—Parent ^{235}U

$$^{223}\text{Ra} \xrightarrow{\alpha} {}^{219}\text{Ra} \xrightarrow{\alpha} {}^{215}\text{Po} \xrightarrow{\alpha} {}^{211}\text{Pb} \xrightarrow{\beta} {}^{211}\text{Bi} \xrightarrow{\alpha} {}^{207}\text{Tl} \xrightarrow{\beta} {}^{207}\text{Pb}$$
$$\text{11.4 d} \quad\quad \text{3.9 s} \quad\quad \text{18 μs} \quad\quad \text{36 min} \quad\quad \text{2.15 min} \quad \text{4.8 min} \quad\quad \text{stable}$$

Thorium Series—Parent ^{232}Th

$$^{228}\text{Ra} \xrightarrow{\beta} {}^{228}\text{Ac} \xrightarrow{\beta} {}^{228}\text{Th} \xrightarrow{\alpha} {}^{224}\text{Ra} \xrightarrow{\alpha} {}^{220}\text{Rn} \xrightarrow{\alpha} {}^{216}\text{Po} \xrightarrow{\alpha} {}^{212}\text{Pb}$$
$$\text{5.8 y} \quad\quad \text{6.1 h} \quad\quad \text{1.9 y} \quad\quad \text{3.6 d} \quad\quad \text{55 s} \quad\quad \text{0.16 s} \quad\quad \text{10.6 h}$$

$$\xrightarrow{\beta} {}^{212}\text{Bi} \xrightarrow{\alpha} {}^{208}\text{Tl} \xrightarrow{\beta} {}^{208}\text{Pb}$$
$$\text{60.6 min} \quad \text{3.1 min} \quad\quad \text{stable}$$
$$^{212}\text{Po}$$
$$\text{0.3 μs}$$

Uranium Series—Parent ^{238}U

$$^{226}\text{Ra} \xrightarrow{\alpha} {}^{222}\text{Rn} \xrightarrow{\alpha} {}^{218}\text{Po} \xrightarrow{\alpha} {}^{214}\text{Pb} \xrightarrow{\beta} {}^{214}\text{Bi} \xrightarrow{\beta} {}^{214}\text{Po} \xrightarrow{\alpha} {}^{210}\text{Pb}$$
$$\text{1600 y} \quad\quad \text{3.8 d} \quad\quad \text{3.05 min} \quad \text{27 min} \quad\quad \text{20 min} \quad\quad \text{160 μs} \quad\quad \text{22 y}$$

$$\xrightarrow{\beta} {}^{210}\text{Bi} \xrightarrow{\beta} {}^{210}\text{Po} \xrightarrow{\alpha} {}^{206}\text{Pb}$$
$$\text{5.0 d} \quad\quad \text{138 d} \quad\quad \text{stable}$$

Fig. 5.1. Decay scheme of principal radium isotopes with halflives of products.

The isotope of particular concern from a health standpoint is ^{226}Ra. This can be distinguished from the other isotopes by the rate at which its α activity changes. The other difference is that ^{226}Ra is the only isotope that produces an α-emitting gaseous product with a reasonable halflife, namely ^{222}Rn (see Fig. 5.1). Hence, if the radioisotopes are chemically separated from the sample, a study of the a activity of this material over a period of a few weeks will enable the composition of the isotopes to be calculated using the halflives shown in Fig. 5.1. Alternatively the isotope mixture can be stored until sufficient radon-226 has been produced to be separated and a-counted.

The radium isotopes are separated by precipitating lead and barium carrier as the sulfates. The precipitate is purified, by washing with nitric acid, dissolved in EDTA and re-precipitated as radium–barium sulfate at pH 4.5. At this pH the other naturally occurring a-emitters and lead are kept in solution.

The procedure used for a water sample[2] is as follows: To 1 litre of sample is added 5 ml of 1 M citric acid, 2.5 ml of concentrated ammonium hydroxide, lead nitrate carrier (4 mg Pb) and barium chloride carrier (6 mg Ba). The barium and lead are then precipitated from the boiling solution with sulfuric acid. The precipitate is separated by decanting and centrifuging. It is then washed with concentrated nitric acid. The precipitate is dissolved in EDTA made alkaline with ammonium hydroxide. Acetic acid is added to bring the pH to 4.5, at which point the barium–radium sulfate precipitates. This precipitation is the point from which the ingrowth of activity is timed. The precipitate is transferred to a counting disc, dried and weighed. The recovery of the barium carrier is calculated to determine the chemical yield. The sample is a-counted to determine the total a-emitting radium isotopes. If it is desired to determine the isotope content by the ingrowth of a activity, the sample is counted periodically for up to about 20 days. From Fig. 5.1 it is seen that ^{223}Ra completes its ingrowth within a few hours of separation and after that the activity decays according to its halflife of 11.4 days. The ingrowth from the first two daughters of ^{224}Ra is complete within a few minutes and the ingrowth of the next a-emitting daughter is governed by the 10.6-h halflife of ^{212}Pb. At the same time the ^{224}Ra is decaying with a halflife of 3.6 days. The ingrowth of ^{226}Ra is governed by the 3.8-day halflife of ^{222}Rn. From these considerations a set of simultaneous equations is set up to determine the contribution of each isotope to the ingrowth curve obtained.

The maximum permissible level for ^{226}Ra in drinking water is a few picocuries per litre. At these levels the statistical problems in counting make it difficult to obtain an ingrowth curve with sufficient accuracy to enable the isotope ratios to be calculated. It is often the practice to obtain the total radium isotopes by the above procedure in order to discover whether it is necessary to measure ^{226}Ra. If the total radium count is low then there can be no problem of undesirable ^{226}Ra levels. If the total radium level approaches that of the limit for ^{226}Ra then the ^{226}Ra can be determined by the radon emanation method.

In this method the radium is separated by precipitation with barium carrier as the sulfate. The precipitate is filtered off and placed in a platinum dish. It is treated with hydrofluoric acid and ammonium sulfate to remove silica and completely solubilize all radium compounds. The sample in the dish is then fumed with phosphoric acid to remove sulfur trioxide. A solution of the sample is made and it is transferred to the radon bubbler tube. The tube is sealed and stored for about three weeks for the ingrowth of ^{226}Rn. After this period the radon produced is swept from the tube, through a drying train, into a scintillation counter. There the a activity of the gas is determined. The equipment used for radon emanation is shown in Fig. 5.2.

If it is required to determine the recovery obtained, the barium carrier used can be ^{133}Ba, which is a γ-emitter. After the radon has been swept from the solution and counted, the solution may be removed and the amount of ^{133}Ba determined by γ counting.

The other environmental samples are treated in a similar way to that used for strontium analysis. Radium, like strontium, is not volatile and the samples can be dry-ashed. The organic samples are ashed and the ash fused with sodium

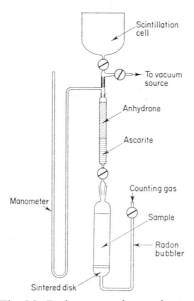

Fig. 5.2. Radon emanation equipment.

carbonate. The alkali metals are leached from the melt with water and the alkaline earths are dissolved in nitric acid. Sediments and soils may be treated the same way by sodium carbonate fusion. For samples and residues which can be dissolved in nitric acid, the alkaline earths are precipitated with phosphoric acid. This precipitate is dissolved in nitric acid. Samples containing large amounts of calcium are treated with fuming nitric acid (as in the strontium method) to remove it. The procedure as for a water sample is then followed.

Cerium

The isotopes of cerium that are found in environmental samples are cerium-141 and cerium-144. These are both β-emitters; they also emit γ radiation. Their half-lives are: ^{141}Ce, 32.5 days; ^{144}Ce, 285 days.

After pretreatment the cerium is extracted into di-2-ethylhexyl phosphoric acid in an *n*-heptane, backwashed into an aqueous solution, and precipitated as the oxalate for counting. Ruthenium contamination is removed by evaporating the backwashed material with perchloric acid.

The procedure for water samples is as follows:[3] to a 1-litre sample, cerium carrier (10 mg Ce) is added and the solution evaporated to 20 ml. This is transferred to a Teflon® centrifuge tube and 5 ml of hydrofluoric acid added. The tube is centrifuged and the supernatant discarded. The precipitate is dissolved in 2 ml of saturated boric acid and 2 ml of nitric acid and diluted with water to 10 ml. Ammonia is then added to reprecipitate the cerium and the tube is centrifuged. The supernatant is discarded. The precipitate is dissolved in 10 ml of a solution of 1 M potassium bromate and 10 M nitric acid. The solution is then extracted with 10 ml of 0.75 M di-2-ethylhexyl phosphoric acid in *n*-heptane. The organic layer is washed and back-extracted into a solution of 2:1 nitric acid containing a few drops of hydrogen peroxide. To the aqueous layer is added a ruthenium carrier (10 mg Ru) and 2 ml of perchloric acid. The solution is evaporated to dryness. The residue is dissolved in 1:1 hydrochloric acid, diluted and the cerium precipitated with oxalic acid. This is then dried, weighed to determine the yield, and counted.

Cerium-144 decays to yield ^{144}Pr which has a halflife of 17.3 min and emits a high-energy β particle (2.98 MeV). With this short halflife the parent and daughter rapidly become in equilibrium and by counting ^{144}Pr the ^{144}Ce can be determined. This gives a means of determining ^{141}Ce and ^{144}Ce since β counting with an absorber counts only ^{144}Pr (which equals ^{144}Ce) and counting without an absorber counts ^{141}Ce, ^{144}Ce and ^{144}Pr. The isotopes may also be distinguished by γ spectrometry. Samples other than water are treated in the same way as for the analysis of strontium (or radium). Cerium, like strontium, is not volatile so that dry-ashing may be used if desired.

Cesium

The two cesium isotopes of concern are cesium-134 and cesium-137. These are both β-emitters and they both also emit γ radiation. Their halflives are: ^{134}Cs, 2.19 years; ^{137}Cs, 30 years.

The sample containing cesium is freed from polyvalent metal ions by precipitating them with sodium hydroxide–sodium carbonate. The alkali metals, Cs, Rb and K, are then extracted from the solution by absorption on ammonium phosphomolybdate (APM). This APM is dissolved and the cesium precipitated with bismuth triiodide. Citric acid is added to prevent the precipitation of the APM. The cesium bismuth triiodide is decomposed by heating with nitric acid and the cesium precipitated as the chloroplatinate for counting.

In the procedure used[4] about 250 ml of sample, plus 40 mg of Cs carrier, is boiled with sodium carbonate–sodium hydroxide. The supernatant liquid is collected, made about 0.5 N with nitric acid and 1 g of ammonium molybdophosphate added. This is stirred for about an hour and then allowed to settle. The precipitate is dissolved in 10 ml of 2 N sodium hydroxide, heated and 5 ml of 30% citric acid added. The solution is evaporated to 5 ml, filtered and the filter washed with a further 5 ml water. The solution is then cooled and to it is added 5 ml bismuth triiodide reagent (40 g BI_3, 40 g KI in 96 ml water and 4 ml acetic acid). The precipitated cesium bismuth triiodide is decomposed by heating with 5 ml of 30% nitric acid. The precipitated iodine is filtered off and the cesium precipitated from the supernatant liquid with chlorplatinic acid. The precipitate is washed with ethanol and weighed for counting. The total radiocesium can be determined by β-counting. The differentiation between ^{134}Cs and ^{137}Cs is best made by γ spectrometry.

Cesium is somewhat volatile and if dry-ashing is used to treat samples the temperature should not be allowed to exceed 450 °C. Sediment and soil samples, and the ash from biological samples, may be fused with sodium carbonate. The fusion mix is leached with dilute hydrochloric acid; any insoluble material remaining is dissolved in 10% hydrochloric acid. The solution is then treated with sodium carbonate–sodium hydroxide as above.

At high levels of cesium ($>1~\mu Ci~l^{-1}$) water samples may be extracted directly with sodium tetraphenylborate in amyl acetate.[5] The organic layer can then be γ-counted directly.

Zirconium

Zirconium-95 is a β-emitter with a halflife of 65 days. It also emits γ radiation. Another zirconium isotope is ^{97}Zr with a halflife of 17 hours; this is not usually found in environmental samples.

Zirconium is extracted from the sample with 2-thenoyl-trifluoroacetone (TTA) in xylene. A back extraction into an aqueous phase is made using a dilute nitric–hydrofluoric acid mixture. This avoids back extracting iodine which is the only interference in the extraction process. The aqueous phase is γ-counted to determine the zirconium-95. The zirconium in the sample is rendered completely soluble by treatment with hydrofluoric acid. Fluoride and sulfate interfere in the extraction process and they are removed by evaporation with perchloric acid. A small amount of carrier Zr is added to avoid problems of adsorption of ^{95}Zn on the vessel walls.

In the procedure used,[5] to the sample is added 1 drop of zirconium carrier solution (10 g l^{-1} Zr) and 2 ml of concentrated hydrofluoric acid. The sample is evaporated to 25 ml and transferred to a dish (platinum or Teflon®). The solution is evaporated to about 5 ml and transferred to a 50-ml centrifuge tube with perchloric acid (1 + 3). A drop of potassium iodide carrier solution (8.3 g l^{-1} KI) is added and the solution evaporated to near dryness. This evaporation with perchloric acid is then repeated. The sample is dissolved in perchloric acid ($\sim$10 ml) and transferred to an extraction tube, where a few drops of 30% hydrogen peroxide are added. The solution is extracted with 10 ml of TTA solution (0.5 M). The organic layer is washed with perchloric acid containing hydrogen peroxide. Back extraction is then made with two 10-ml portions of hydrofluoric–nitric acid solution. The aqueous layer is γ-counted to determine the zirconium-95.

Biological samples are ashed and the residue fused with sodium carbonate. Sediment and soil samples may also be fused. The fusion mixture is dissolved in water to extract the alkali metals and the residue dissolved with hydrofluoric acid and treated as above.

Iodine

Iodine-131 is a β-emitter with a halflife of 8.07 days. It also emits γ radiation. There are other active iodine isotopes besides ^{131}I but they are short-lived and in most environmental samples ^{131}I is the only isotope of concern. To determine radioactive iodine the sample solution is passed through a cation exchange resin which extracts the active cations. The iodine in solution then undergoes isotope exchange with silver iodide and the resulting active silver iodide is counted. To ensure that all the iodine in the sample is present as iodide an oxidation–reduction cycle is performed. The oxidation is carried out with alkaline hypochlorite; the reduction is carried out using bisulfite in an acid medium. In the procedure used[6] the sample solution of 1 litre or less is made alkaline with sodium carbonate and 25 ml of sodium hypochlorite solution added (5.25% NaOCl—household bleach solution). The solution is boiled and then acidified. The chlorine is removed by boiling and 5 ml of a 20% solution of sodium hydrogen sulfite then added. This solution on cooling is run through a column of ion exchange resin, phenol–sulfonic acid type. The effluent from the column is added to a beaker containing a precipitate of silver iodide and a small amount of sodium fluoride. The precipitated silver iodide and the column effluent are mixed together for about 15 min, after which the silver iodide is filtered off and mounted for counting. The counting may be either β-counting in a proportional counter or it may be γ-counting using a multi-channel analyser.

Other samples may be treated by the same technique. Solid samples may be directly analysed by a γ-spectrometer. Iodine may be present in air either as a particulate or in gaseous form; the filter or adsorbent used to collect it may be extracted to solubilize the iodine or they may be analysed by γ spectrometry.

SOURCES OF RADIONUCLIDES

The radionuclides that enter the environment from man's activities do so in several ways. The principal sources of these materials are:

1. Weapons testing
2. Nuclear power generation—both in emission in day-to-day operation and in processing the spent fuels from the reactor
3. Mining and processing of radioactive minerals.

When radioactive materials are found it is often desirable to be able to make some estimate of their probable source. With the ability to determine different nuclides of the same element by such techniques as γ-spectrometry it is possible to distinguish between these various sources of radioactivity.

In the case of material produced by fission, either in a weapon or in a nuclear reactor, the relative abundance of the various nuclides formed is known. The short-lived nuclides disappear at faster rate than the long-lived ones. Hence a measurement of the ratios of ^{89}Sr and ^{90}Sr with halflives of 52 days and 28.1 years respectively, or ^{141}Ce and ^{144}Ce with halflives of 32.5 days and 285 days respectively, is a measure of the time elapsed since their production. This gives some indication whether the radioactivity is due to 'old' bomb fall-out or due to material recently produced in a reactor.

Some nuclides are formed primarily in long irradiations rather than by fission. ^{134}Cs, for example, is produced by (n, γ) reaction with the 'normal' ^{133}Cs. Hence a determination of ^{134}Cs and ^{137}Cs indicates whether the activity is from a nuclear reactor or a weapon. In the case of heavy water moderated cooled reactors, the main activity released in day-to-day operation is tritium with some ^{131}I.

The mining and processing of radioactive minerals releases uranium, thorium and radium. The relative abundance of these nuclides is different in some of the ore bodies and a determination of these can then indicate where these materials originated. Similarly the ratio of ^{235}U to ^{238}U in natural uranium is known and this provides a means of determining whether any uranium found is from a natural or an enriched uranium source.

REFERENCES

1. D. C. Camp, C. Gatrousis and L. A. Maynard, *Nucl. Instrum. Methods* **117**, 189 (1974).
2. *Standard Methods for the Examination of Water and Wastewater* 14th Edn, Part 700, APHA-AWWA-WPCF (1975).
3. *HASL Procedure Manual* (Ed. John H. Harley), HASL 300, ERDA Health and Safety Laboratory, New York, 1973.
4. A. Mirna, *Analyst* **95**, 1000 (1970).
5. *ASTM Standards*, Part 31, American Society for Testing and Materials, Philadelphia, Pa. (1976).
6. W. J. Maeck and J. E. Rein, *Anal. Chem.* **32**, 1079 (1960).

CHAPTER 6

THE DETERMINATION AND IDENTIFICATION OF ORGANIC MATERIALS

There are two concerns in considering the determination of organic materials in the environment. One is the potential of the organic material as a source of carbon for degradation by microorganisms and the resulting effect on water quality. The other is the effect of specific organic chemicals on plant and animal life.

The first of these concerns arises in the following manner: almost all the organic materials that enter natural waters will be degraded by the micro-organisms in the water, the end-products of this degradation being carbon dioxide and water. The oxygen dissolved in the water is used to carry out this oxidization and in the process it becomes depleted. If there is an excessive amount of organic material, then the oxygen may become depleted to the extent that not enough remains to support the animal life in the water. Hence the discharge of organic material to a receiving water may result in the water being unfit for fish, not because of any direct toxic effect of the material but because of the depletion of the oxygen resulting from the material's metabolization by microorganisms. This same effect occurs in the eutrophication of a water body. An abundance of nutrients in the water causes a prolific growth of algae in the summer months. In the winter these algae die and fall to the bottom where they are degraded by bacteria. The resulting depletion of oxygen causes first of all a change in the fish population. The game fish such as trout, which require a high dissolved oxygen level to survive, disappear and are replaced by fish, such as catfish and carp, which can live with lower dissolved oxygen levels. Eventually, as the process proceeds, the lower levels of the water may become completely anoxic. At this stage the process becomes self-perpetuating since the nutrients, such as phosphorus, under anoxic conditions can become mobilized from the sedimented material. They are then available to participate in another cycle of abundant plant growth followed by decomposition of the plant material and oxygen depletion.

Many organic chemicals have an effect on the plants or animals by which they are metabolized. In some cases there is progressive accumulation in the biota so that the final consumer in the food chain ingests levels of the materials that are many thousands of times those present in the water or soil. Perhaps the best known example is that of the chlorinated hydrocarbons such as DDT. These accumulate in the food chain to such an extent that the birds at the end of the chain receive an amount large enough to disturb their reproductive-eggshell system. With the similar chlorinated biphenyls there is concentration to the extent that fish in some waters are unfit to eat, although the biphenyls in the water are at the 'parts per trillion' level. In these cases it is necessary to be able to identify and quantify the specific materials, including isomers. This is generally done by the immensely powerful separation processes available with chromatographic techniques. These are described at the end of the chapter.

Some organic materials present problems because of their chemical behaviour. For example, phenols are known to cause difficulties in drinking water supplies because they form the bad-tasting chlorophenols when the water is chlorinated. In this case the determination carried out is a generic one for a 'phenol'; the individual phenols historically have not been identified.

Hence the determinations carried out range from the completely non-specific 'organic carbon' measurement, through the determination of classes of compounds, to the specific identification of a single organic compound. The procedures used are described in that order.

ORGANIC CARBON

In the first two determinations described, biochemical and chemical oxygen demand (BOD and COD), the equivalence of the sample in terms of its reaction with oxidizing agents is measured. In the other determination the organic content is measured more or less directly.

Biochemical oxygen demand

In the determination of the biochemical oxygen demand (BOD_5) the sample is incubated at 20 °C for five days. The dissolved oxygen is measured at the beginning and the end of the period. The difference represents the amount of oxygen used in the biochemical oxidation of the sample. Although the oxidation is usually not complete, five days is used for the incubation period as it has been found that in this time a reasonably large fraction of the oxidation occurs. The test is an empirical one that has been used for many years, the basis for its use being that the consumption of oxygen in the incubator bears some relation to the organic load experienced in the receiving waters. There has been much discussion as to whether or not this is true, and in the ASTM Standards since the 1971 edition the BOD_5 test procedure for industrial water and industrial waste waters has been omitted. This was done on the grounds that the test is difficult to standardize, that the degradation pattern occurring in the laboratory bottle does not necessarily match that which will occur in the receiving waters, and that other methods of measuring the organic loading are more valid. However, the BOD_5 test does at least attempt to consider the biodegradation of the material in the receiving waters, which is fundamental to the issue as to whether the discharge does represent an oxygen-consuming load, and the test continues to be in widespread use. It is included in the 14th edition (1975) of *Standard Methods*.[1]

The procedure used depends upon the type of sample. Since the oxygen for the oxidation must come from the dissolved oxygen in the incubated sample (the saturation value is about 9 mg l^{-1} O_2 at 20 °C) a sample having a BOD greater than about 7 mg l^{-1} must be diluted. The sample is aerated before incubation to ensure that it is saturated with oxygen. In some samples taken during the winter months, or taken when algae are actively growing, the dissolved oxygen level is above the saturation level at 20 °C. In these cases the level is brought down to the saturation level by shaking the sample in a half-filled bottle or by aerating it. The basis for the test is that there are microorganisms present which will degrade the organic material. Where the sample is a domestic waste water or a surface water there are probably sufficient bacteria present for this. However, where the effluent has been chlorinated, or has been subjected to high temperature, it is generally necessary to add a 'seed', i.e. a small amount of a culture containing microorganisms that will degrade the organic materials. Where the effluents are normal domestic effluents a seed of domestic wastewater may be used. In the case of industrial effluents, however, it is generally necessary to use a seed from a culture that has been acclimatized to feed on the materials being discharged.

The sample is pretreated before the test to remove materials that would kill the microorganisms. Chlorine is removed with sodium sulfite, and the sample is brought to a pH of about 7 with either sodium hydroxide or sulfuric acid. A buffer solution containing phosphate and ammonium chloride is added to the sample, together with solutions containing magnesium, calcium and iron, in order to ensure that there are 'inorganic' nutrients for the growth of the microorganisms.

The sample containing these nutrients, seeded and diluted if necessary, is placed in a container in the incubator. Typically a 'BOD bottle' is used which is a bottle about 250 ml in capacity with a ground-glass stopper. There is a rim around the top of the bottle neck so that, when the stopper is in place, a water seal can be made around the top of the stopper. The bottle is filled completely so that there is no air space. The dissolved oxygen is measured before the sample is placed in the BOD bottle and after it has been incubated for five days.

The dissolved oxygen measurement is usually carried out using a membrane electrode.[1a] This is an electrode, of either a polarographic or a galvanic type, which is covered with a membrane permeable to oxygen. This is immersed in the sample solution to determine the dissolved oxygen. The measurement can also be made chemically. Traditionally this has been done by the 'Winkler' method or by one of its modifications. In this method a solution of manganous sulfate and sodium hydroxide is added to the sample. The precipitated manganous hydroxide is rapidly oxidized by the dissolved oxygen to manganic hydroxide. This, in acid solution, is then used to liberate iodine from potassium iodide. This iodine may be determined spectrophotometrically or by titration with thiosulfate. The interference most commonly found is nitrite which also oxidizes the manganous hydroxide. This interference is overcome by adding sodium azide to the sodium hydroxide solution used to precipitate the manganous hydroxide.

No preservatives are added to water samples collected for the BOD determination. The samples should be cooled to about 4 °C and the incubation should begin within 24 hours of the sample being collected.

Chemical oxygen demand

In the determination of the chemical oxygen demand (COD) the sample is heated with an acid solution of potassium dichromate. The oxidizable matter in the sample reacts with the dichromate and the excess dichromate is then determined by titration with ferrous ammonium sulfate. Most types of organic material are oxidized in this treatment; however, straight chain aliphatic compounds, aromatic hydrocarbons and pyridine are not. To obtain oxidation of the straight chain aliphatic compounds, such as the alcohols and acids, silver sulfate is used as a catalyst. This gives some problems in the presence of chloride due to the precipitation of silver chloride. To overcome this precipitation, mercuric sulfate is added to convert the chloride to a soluble mercuric chloride complex.

In the procedure used for a manual determination[1] 50 ml of sample is placed in a 500-ml flask and 1 g of mercuric sulfate and 5 ml of sulfuric acid reagent are added. The sulfuric acid reagent is made by dissolving 22 g of silver sulfate in one 4-kg bottle of concentrated sulfuric acid, 25 ml of 0.25 N potassium dichromate solution is added and the mixture is refluxed for two hours. The contents of the flask are cooled and diluted. The excess dichromate is then determined by titration with 0.01 N ferrous ammonium sulfate solution using ferroin as an indicator. A blank of distilled water is refluxed and titrated by the same procedure. This procedure is used for samples with a COD greater than about 50 mg l^{-1}; for samples with lower levels, 0.025 N dichromate solution is used. Nitrite exhibits a COD; however, because of the comparatively low levels of nitrite found in most water samples (a maximum of 1 or 2 mg l^{-1} N in even the most grossly polluted water) this is not normally considered a significant interference. It may be removed, if desired, by the addition of sulfamic acid to the refluxing flask. The procedure may also be carried out in Auto Analyzer® equipment.

The technique of COD determination has some advantages over the determination of BOD, particularly in that the five-day incubation period is not required. The reproducibility of determination with particular waste materials is also better, although it must be recognized that not all of the organic matter is oxidized in the procedure. However, it gives no measure as to whether the organic materials that are oxidized chemically will be biodegraded in the receiving waters.

Samples for COD determination may be preserved by the addition of sulfuric acid. Samples containing solid materials may be analysed by homogenizing them before the aliquot is taken for addition to the refluxing flask.

Adsorbable organic carbon

In this technique the water to be examined is passed through a bed of activated carbon. The organic materials are adsorbed on the carbon and are subsequently eluted with an organic solvent. The organic extract is evaporated and the residual material, representing the organic material in the water, is determined gravimetrically. The organic solvent commonly used is chloroform and the test is described as the carbon–chloroform extract (CCE). This type of determination has been developed for the examination of waters containing levels of organic carbon of a few mg l^{-1} C, which may be used as sources of drinking water. Historically there has been difficulty in measuring organic carbon at these levels by other techniques that do not involve complicated and expensive equipment.

A miniaturized version of the text equipment has been developed and is described in *Standard Methods*.[1] This is designated CCE-m, the 'm' indicating the use of the miniaturized sampler. The activated carbon (70 g) is contained in a piece of 2-in. PVC pipe 3 in. long and the water flows through it at a rate of about 20 ml min^{-1} for a 48-hour period. Every 30 min the flow rate is increased to about 400 ml min^{-1} for 7–8 s, to provide a fast forward flush that removes both carbon fines and any entrapped air from the adsorption bed. After the sampling period the activated carbon is removed and dried at 40 °C. The dried carbon is extracted with chloroform in a soxhlet extractor. The extract is reduced in volume by heating and finally evaporated to dryness at room temperature in a stream of air. The dried extract is then weighed to determine the weight of organics.

The results obtained with this technique range from about 0.1 mg l^{-1} CCE-m with clean water to about 1.2 mg l^{-1} CCE-m obtained with heavily polluted water. In a study of the process[2], using a dissolved organic–carbon monitor to measure the influent and the effluent water, it was found that the adsorption of organics on the carbon had an efficiency of 70–90%. However, the extraction efficiency, determined by elemental analysis on the chloroform extract, was only about 15%.

Total organic carbon

The organic carbon content of a sample may be determined by oxidizing the organic materials to carbon dioxide, separating the carbon dioxide from the sample mixture and measuring it. The choices in the method are the way in which the oxidation is carried out and the way in which the carbon dioxide is measured. In the oxidation process the choice is one of gas-phase oxidation or wet oxidation. In the gas-phase oxidation process the oxidation is carried out by passing the vaporized sample, mixed with oxygen, over a catalyst at high temperature. In the procedure described by Van Hall et al.,[3] which is used, for example, in the Beckman Total Carbon Analyser, the sample is passed over a cobalt oxide catalyst at 950 °C. In this case the carbon dioxide in the effluent gas is measured with a non-dispersive infrared analyser. In this procedure the sample (up to about 200 μl) is injected into a stream of oxygen with a syringe. The vaporized sample passes through the catalyst contained in a tube in the furnace. The resulting carbon dioxide is then measured with the infrared analyser. A schematic of this equipment is shown in Fig. 6.1. The method will measure carbon levels down to about 0.5 mg l^{-1} C. There are a number of automated versions of this technique available.

The measurement of the carbon dioxide resulting from the oxidation of the carbon suffers from the interference from inorganic carbonates which also give carbon dioxide under these conditions. The carbonates in the sample may be decomposed by acidification and the resulting carbon dioxide swept from the

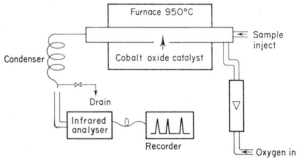

Fig. 6.1. Schematic of carbon analyser using gas-phase oxidation.

solution with a stream of gas before the oxidation is carried out. This has the disadvantage that volatile organic materials may be lost from the sample. Another option is to determine that total carbon, i.e. the carbon dioxide from the carbonates plus that formed in the oxidation, and then separately measure the carbon dioxide produced by the acidification of the carbonates. The difference represents the organic carbon. These two determinations can be carried out in versions of the analyser that are equipped with two channels; one contains a conventional oxidation furnace for the 'total' determination, the other a tube packed with acid packing and operated at about 150 °C, where the carbonates in the sample are decomposed. The technique of determining organic carbon by the difference between 'total' and 'inorganic' has the disadvantage that the precision of the organic carbon determination is poor when the inorganic carbon levels are much greater than the organic carbon levels.[4] When used to measure the total organic carbon in a sample containing suspended solids the sample must be well homogenized before the aliquot is taken into the syringe. The small aliquot taken and the narrow-bore needle used in the syringe give some problems in obtaining a representative aliquot of such samples.

In wet oxidation techniques a larger sample than in the gas-phase oxidation methods is usually taken. This results in a higher sensitivity being obtained, so that levels down to about 10 μg l^{-1} C can be measured. The larger sample size also makes it easier to obtain a representative sample. Probably the most commonly used procedure is that described by Menzel and Vaccaro.[5] In this, the sample (10 ml) is placed in a glass ampoule, acidified with phosphoric acid and the carbon dioxide from the inorganic carbon swept from the solution. Potassium peroxydisulfate (0.4 g) is added and the ampoule is sealed. The ampoule is then autoclaved at 130 °C and 30 psig for 30 min. After the ampoule is cooled and opened the carbon dioxide is measured with a non-dispersive infrared analyser. This process is available as a commercially packaged instrument. Williams[6] used this technique to study the measurement of organic carbon in sea water, using ^{14}C to follow the oxidation. He found that the completeness of oxidation was higher at 100 °C than at 130 °C. Baldwin and McAtee[7] showed that, in a variety of natural water samples, complete oxidation could be obtained by treating the sample in the ampoule with silver catalysed peroxydisulfate at room temperature.

The wet oxidation techniques are amenable to automation. Erhardt[8] describes an automated system where a sea water sample is oxidized by irradiation with ultraviolet light in the presence of peroxydisulfate. The carbon dioxide is then measured by determining the change in conductivity when it is absorbed in sodium hydroxide solution. An automated system using an infrared analyser for determining the carbon dioxide has been described.[9] Here the oxidation is carried out using either ultraviolet irradiation or silver-catalysed peroxydisulfate.

In both this method and the one of Erhardt[8] a 1200 W low-pressure mercury lamp is used as the ultraviolet source. This requires a cooling ventilation system of considerable capacity to remove the heat and ozone given off. An automated system that uses a low power germicidal lamp (about 45 W) to provide the ultraviolet irradiation for oxidation has been developed by Technicon Instruments Corporation.[10] In this system the carbon dioxide produced by the oxidation is separated from the sample matrix by passage through a dialyser containing a silicone rubber gas-permeable membrane. The receiving stream is a weakly buffered solution of carbonate–bicarbonate containing phenolphthalein. The change in colour as the pH changes by the absorption of the carbon dioxide is measured with a colorimeter.

An alternative means of measuring the carbon dioxide produced by the oxidation of organic carbon is to reduce it catalytically to methane, and determine the methane with a flame-ionization detector. This is a technique described by Takahashi et al.[11] and is used in some of the commercial instruments for carbon analysis.

The wet oxidation techniques suffer from the same problem as is discussed in the description of the gas-phase oxidation process, namely the difficulty of distinguishing between the carbon dioxide from the oxidation of the organic carbon and that released from inorganic carbonate. In those wet oxidation processes that remove the inorganic carbonate by acidification and gas stripping, the problem is more serious because of the higher sensitivity of the detection method. Any inorganic carbon remaining in the solution after the gas stripping will be measured as apparent organic carbon. The removal of inorganic carbonates from samples has been discussed by Van Hall.[12]

Determination of organic carbon in solid samples may be carried out in an 'elemental analyser' where the sample is combusted and the resulting gases separated by chromatographic techniques and measured. Typically the elemental analyser will measure C, H and N.

Water samples in which it is desired to measure only the organic carbon may be preserved by acidification and low temperature storage. If inorganic carbon is also to be determined the sample must not be acidified; low temperature storage with prompt analysis is the only preservation technique available.

SURFACTANTS

The two types of surfactants usually of concern in environmental samples are anionic and non-ionic. They are the only two types which are used to any extent in the formulation of domestic and industrial detergents. These detergents are the main source of surfactants released to the environment. The bulk of the surfactants used in the past have been anionics; however, nowadays non-ionics constitute a significant proportion of the surfactants used.

Anionic surfactants

The most common anionic surfactant used is that made by the sulfonation of an alkyl benzene. In the early days of heavy-duty detergent manufacture the alkyl group attached to the benzene ring was mainly a branched hydrocarbon chain about twelve carbon atoms long. It was found that this material (then termed alkyl benzene sulfonate (ABS)) was not readily biodegraded in treatment plants and receiving waters. It was shown that the quarternary carbon at the branching of the side chain was the terminal point for the biodegradation that took place. The use of a straight chain was adopted in order to produce a material that is more readily degraded. The benzene ring may be attached at various points in the hydrocarbon chain. To distinguish this material from the ABS, it was called linear alkylate sulfonate (LAS). Other synthetic anionic types used are alkyl sulfates made by the sulfation of fatty alcohols or of fatty alcohol–ethylene oxide condensates. When determining the anionic surfactants in water samples the total is determined and often referred to as 'LAS', although any other anionic surfactant present will also contribute to the measurement made. If desired, a distinction between alkyl sulfates and sulfonates can be made by boiling the sample with acid. Alkyl sulfates hydrolyse under these conditions, sulfonates do not. Usually the alkyl sulfates and other surfactants such as soaps do not constitute a large fraction of the anionic surfactants present since they are readily degraded.

The traditional method of analysis depends on the fact that methylene blue will react to form a salt with anionic surfactants. This salt is soluble in chloroform and is extracted from the aqueous solution. The absorbance of the chloroform extract at 652 nm is used to measure the concentration of the surfactant. The analysis is described as the determination of the methylene-blue-active substances (MBAS). The manual procedure is described in *Standard Methods.*[1]

In the procedure a calibration curve is prepared using LAS as the standard material, and the MBAS concentration is reported in terms of 'apparent LAS concentration'. The sample (up to 400 ml) is placed in a separating funnel and made alkaline; methylene blue reagent is added and the sample is extracted three times with 10-ml portions of chloroform. The chloroform extract is washed and the absorbance at 652 nm measured. The method can also be carried out in Auto Analyzer® equipment. Substances, other than anionic surfactants, which complex methylene blue give positive interferences, as do inorganic ions such as cyanates, chlorides, nitrates and thiocyanates, which form ion-pairs with methylene blue. High salt content in the sample also leads to a salting out of methylene blue ion-pairs into the chloroform layer. The method will determine MBAS down to about 0.025 mg l^{-1} apparent LAS. An alternative method, described in *Standard Methods*,[1] uses adsorption on activated carbon to separate LAS from the water. The adsorbed LAS is then extracted with methanol and the 1-methylheptylamine complex made. This is then determined by infrared absorption measurements.

A titrimetric method suitable for waters of high salinity is described by Wang et al.[13] In this procedure the sample was treated with an excess of a cationic reagent, cetyldimethylbenzyl ammonium chloride. The excess of the cationic reagent was then determined by titration with sodium tetraphenylboron. The sample was first treated with the cationic reagent, chloroform was added and the pH brought to 3.0. At this pH the excess cationic reagent was partially soluble in the chloroform layer to give a yellow colour. As the titration with the tetraphenylboron proceeded the cationic reagent was removed from the aqueous solution. This resulted in the disappearance of the yellow colour from the chloroform layer.

A titrimetric method using a surfactant extractive electrode is described by Ciocan and Anghel.[14] Using the electrode as the end-point detector in a titration of two different anionic surfactants with 0.004 M cetyldimethylbenzyl ammonium chloride, they were able to distinguish the end-point due to each surfactant.

Rather than using a wholly organic cationic reagent to determine the anionic material, cationic metal complexes can be used. Taylor and Waters[15] describe a radiometric method in which the anionic material is extracted into a solvent as an ion-pair with tris(1,10-phenanthroline) iron(II) labelled with iron-59. This method has a detection limit of 5 µg l^{-1} LAS. Crisp et al.[16] have extracted the anionic surfactants into chloroform as an ion-associated compound with the bis(ethylenediamine) copper(II) cation. This solution was then measured for copper using colorimetry or flame atomic absorption spectrometry. The detection limits obtained were 30 μg l^{-1} and 60 μg l^{-1} LAS respectively. In a method reported later,[16] the ion-associated compound was extracted from a 750-ml sample and the determination of copper was carried out using AAS with a graphite furnace. The detection limit obtained was 2 μg l^{-1} LAS.

Non-ionic surfactants

The non-ionic surfactants in common use are those made by the condensation of ethylene oxide with either an alcohol or an alkyl phenol. These condensations result in compounds containing ethylene oxide chains containing up to 60 ethylene oxide units; the normal range for detergent use is about 10 units. Propylene oxide is sometimes used in place of ethylene oxide. In the 'generic' determinations described below, the standards used to obtain a calibration are generally made from a commercial surfactant and the results obtained are reported as apparent levels of this surfactant. In the ethylene oxide condensation reaction the overall ratio of alcohol or alkylphenol to ethylene oxide that is used is known, and this is reported as the nominal composition, e.g. R(EO)$_9$. However, the ethylene oxide additions proceed one at a time with almost equal probability, so that the number of ethylene oxide units in the molecules in fact has a probability distribution centred about 9.

The formation of a blue complex between the surfactant and ammonium cobaltithiocyanate was used by Greff et al.[17] The complex was extracted from the aqueous solution, saturated with sodium chloride, into benzene. The absorption of the benzene layer at 320 nm was used to determine the amount of surfactant present. The method had a detection limit of 0.1 mg l^{-1}. This method was recently re-examined by Nozawa et al.[18] Baleux[19] used the reaction between the surfactant and potassium iodide for a colorimetric determination. In this procedure an iodine–potassium iodide reagent was added to the sample and the absorption of the aqueous solution at 500 nm determined. This method had a detection limit of about 0.5 mg l^{-1}. In the colorimetric method described by Favretto and Tunis[20] the surfactant was reacted with a sodium nitrate–picrate reagent to form a picrate complex. This was extracted into 1,2-dichloroethane and the absorbance at 378 nm measured to determine the concentration. This method was used to measure levels of surfactant in the 0.1–1.0 mg l^{-1} range.

Probably the most widely used method is that described by Wickbold.[21] In this the surfactant is concentrated by blowing air, saturated with ethyl acetate, through the sample. The surfactant partitions into the ethyl acetate. The ethyl acetate is evaporated to dryness and the surfactant dissolved in water. This solution is then treated with modified Dragendorff reagent (KBiI$_4$–BaCl$_2$–glacial acetic acid) when a surfactant Ba(BiI$_4$)$_2$ complex precipitates. This is filtered off, washed with glacial acetic acid and dissolved in ammonium tartrate solution. The bismuth concentration is determined by a potentiometric titration with pyrrolidinedithiocarbamate solution, and this is a measure of the surfactant concentration in the sample. The method will measure levels down to about 20 µg l[-1] non-ionic surfactant.

The methods above all rely on surfactant properties for the determination of the materials. It is often desirable to identify the particular material by the presence of either alcohol or alkylphenol groups and by the number of ethylene oxide chains. This is a complex process because, as described above, a typical commercial material contains many chains of ethylene oxide of many lengths and, in addition, the starting alcohol or alkyl phenol is generally a mixture of compounds. When the concern in analysis is to determine the extent of degradation of these materials in the treatment system or receiving waters, it is necessary to identify the degradation products as distinct from following the loss of surfactant properties. It may be that the materials will lose their surfactant properties by the degradation of, for example, the alcohol part of the molecule, leaving the rest of the molecule intact. Hence the loss of surfactant properties does not necessarily mean the complete biodegradation of the material. Chromatographic techniques have been used to determine the surfactants themselves and also to identify the intermediates in the degradation process. Liquid chromatography using a molecular sieve was used by Cassidy and Niro[22] to analyse polyoxethylene surfactants and their degradation products in industrial process waters. Stancher et al.[23] have extracted the non-ionics from water with 1,2-dichloroethane and examined the extracts by gas–liquid chromatography.

In a study of the biodegradation of a linear alcohol–polyethylene oxide surfactant, Tobin et al.[24] described a technique in which the surfactant was cleaved at the ether linkage with hydrogen bromide. The resulting alkyl bromides were then determined by gas–liquid chromatography. This technique had the advantage that the mixture of products obtained comprised the different alcohol chain lengths and the different ethoxylate chain lengths, but not the innumerable possible combinations of them together.

A review of the methods of analysis of detergent products has been made by Longman.[25]

PHENOLS

Phenols are of particular concern in waters used for drinking water supplies since on chlorination they may produce the odoriferous and bad-tasting chlorophenols. Phenols are defined as the hydroxy derivatives of benzene and its condensed nuclei. They may be determined as a class by colorimetric or fluorimetric methods. Individual phenols can be separated and identified by chromatographic techniques; gas–liquid, and high speed liquid, chromatography are commonly used.

The most widely used colorimetric determination is that involving oxidative coupling with 4-aminoantipyrine (4AAP). A manual procedure for this is described in *Standard Methods*.[1] There are a number of potential interfering materials in natural waters and effluents and the phenols are separated from them by distillation. The phenols are not appreciably concentrated in the distillation and in the procedure a 500-ml sample is taken, brought to pH 4 and distilled in the presence of copper sulfate (which prevents the distillation of sulfur as hydrogen sulfide). The distillate is adjusted to pH 10.0 ± 0.2, and 4AAP solution and an oxidizing agent are added. The oxidizing agent may be potassium ferricyanide or peroxydisulfate solution. If the concentration of phenol is >1 mg l^{-1} the absorbance of the aqueous solution can be measured at 510 nm. For lower concentrations, the colour is extracted into chloroform and the absorbance measured at 460 nm. In some cases it may be necessary to carry out a second distillation to obtain a clean distillate. If this second distillate is turbid, the sample may be 'cleaned' up by extracting the phenol from the acidified sample with chloroform or ether, back-washing the solvent with alkali and then carrying out the distillation.

Not all the phenols react with 4AAP; in particular, the method does not determine the *para*-substituted phenols in which the substituent is an alkyl, aryl, nitro, benzoyl, nitroso or aldehyde group. Hence the commonly occurring *p*-cresol is not determined by this technique. The differently substituted phenols react to form compounds with different absorption maxima and different extinction coefficients from phenol itself. Phenol (C_6H_5OH) is used to prepare the calibration curve and the concentrations are calculated and reported as apparent phenol. The detection limit of the solvent extraction method is about 1 µg l^{-1} phenol. The pH in the colour-forming step has an effect on the reaction. To determine halogenated phenols a pH of 7.9–0.1 is used; 2,4-dichlorophenol is used to obtain the calibration curve.

Phenols also react with 3-methyl-1,2-benzothiazolinone hydrazone (MBTH) in an oxidative coupling reaction to form a blue colour. The reaction is more sensitive than that with 4AAP, and MBTH reacts with many of the *para*-substituted phenols which do not react with 4AAP. An automated method involving distillation and reaction with MBTH has been described for the determination of phenol in water.[26]

Phenols may also be determined by a fluorimetric method. In acid solution phenols fluoresce at about 305 nm when excited with radiation at about 275 nm.

An alternative to distillation for separation of phenols from the sample matrix is solvent extraction. An acidified sample is extracted with solvent and the phenols can be back-washed into the aqueous phase under alkaline conditions. A variety of solvents have been used. In a study of the various methods of determining phenols in water, Afghan *et al.*[27] recommended the use of butyl acetate or isoamyl acetate.

The individual phenols may be separated by chromatographic techniques. In *Standard Methods*[1] a gas–liquid chromatographic method is described which uses direct injection of a few µl of the aqueous sample. In the analysis of plant tissue the phenols are often derivatized to improve volatility. In the procedure described by Drawert and Leupold,[28] the trimethylsilyl derivatives are prepared for examination by gas chromatography. The pentafluorobenzyl ethers have also been used for GLC analysis. The determination of pentachlorophenol (PCP) has been described by Chau and Coburn.[29] The PCP was extracted from the sample with benzene; the acidic components, including PCP, were then back-extracted into 0.1 M potassium carbonate solution. The acetate derivatives were prepared by treatment with acetic anhydride and extracted into hexane. The hexane solution was then injected into the gas chromatograph. The detection limit was 0.01 µg l^{-1}.

The lampricide, 3-trifluoromethyl-4-nitrophenol (TFM), has also been determined by a similar technique.[30] The TFM was extracted from the water sample with the macroreticular resin XAD-7, eluted from the resin with ethyl ether and acetylated. The acetate derivative was extracted into benzene and determined by g.l.c. The detection limit was 0.01 µg l^{-1}.

Liquid chromatography is also used to separate phenols in water. In the method described by Wolkoff and Larose[31] the separated phenols in the stream from the chromatograph were detected using a cerium(IV) sulfate fluorescence system. The phenols reduce the Ce(IV) to Ce(III) and this change is measured by a fluorescence detector. Levels down to about 0.4 µg l^{-1} phenol were measured by this method. Thin-layer chromatography is also used.[32]

Phenols in sediment and other solid samples may be determined by carrying out a steam distillation to separate them. The phenols may also be separated from the matrix by solvent extraction.

Preservation of water samples for phenol analysis is by acidification to a pH of about 4 and the addition of 1 g of copper sulfate per litre of sample. Bismuth nitrate may be used.[27] This serves like copper to precipitate interfering sulfur compounds. Oxidizing agents in the sample must be removed and this is generally carried out by the addition of ferrous sulfate or sodium arsenite.

TANNINS AND LIGNINS

Tannins and lignins are natural constituents of plant material and are found in surface water as the result of biodegradation of these materials. They are commonly found in industrial effluents, particularly those from the paper and the tanning industries. Tannins are also added to boiler water to facilitate the handling of calcium sludge deposits and to prevent the formation of hard scale inside the boiler. They are mixtures of complex hydroxylated aromatic compounds. They are determined by their reaction with tungstophosphoric and molybdophosphoric acids. Acting as reducing agents they produce a blue colour with these mixtures; the absorbance at about 700 nm is used to measure the blue colour produced. Any other reducing agents will produce the same colour so that the reaction is completely non-specific. Standard solutions of tannic acid are used to obtain a calibration curve; the results are reported as mg l^{-1} tannic acid, or, multiplied by 2.5, reported as a lignin number. The detection limit is about 1 mg l^{-1} tannic acid. In the Kraft process for producing paper pulp the wood material is treated with sodium sulfite. This treatment results in the production of lignosulfonates. These materials are discharged as waste and it is often necessary to measure the concentration of lignosulfonate waste, or spent sulfite liquor (SSL) as it is often called. The Pearl–Benson method is often used for this. This involves reaction with nitrous acid, which forms nitrosophenols with the phenol groups in the lignosulfonate. The nitrosophenols rearrange in alkaline solution to form coloured quinine oximes. The absorbance at 430 nm is measured to determine the concentration. This reaction proceeds with any phenol, including of course the tannins and lignins. This determination is used mainly to trace the dispersion of sulfite–liquor waste in the receiving waters, using the discharged spent sulfite liquor to prepare the standards.

The lignosulfonates can also be determined by fluorescence. Wilander *et al.*[33] describe a method in which the lignosulfonates are determined in natural waters using an excitation wavelength of 285 nm and an emission wavelength of 405 nm. The method was optimized to reduce the fluorescence obtained from other naturally occurring materials such as humic substances. In a study of the composition of the organic materials in the River Rhine, Eberle *et al.*[34] used extraction to separate the lignosulfonates. The water was first extracted with chloroform at pH 4 to remove 'regular' phenols. An extraction with trioctyl-amine in chloroform at pH 4 was then made to extract the lignosulfonates and the humic substances. The chloroform extract was then back-washed with alkali to bring the organic materials into aqueous solution. This was treated with nitrous acid to form the nitrosophenol derivatives of the lignosulfonates and humic substances. A differentiation of these materials was then made by polaro-graphic measurements.

Chemical preservatives are not used in water samples which are to be analysed for the tannin/lignin type materials. Low temperature storage and prompt analysis are the only preservative techniques available.

NITRILOTRIACETATES

Nitrilotriacetic acid (NTA) and its salts may be determined using chromato-graphic techniques. NTA has been used as a replacement for polyphosphates in detergents and almost the entire concern for NTA in the environment has been about its effect when it is discharged, via sewage effluents, to the receiving waters. Hence the analytical methodology has been developed for its analysis in sewage effluents and water samples. NTA is a complexing agent and a large number of the analytical techniques use this complexing property as the basis for its measurement.

One of the earlier routine measurements was that described by Thompson and Duthie.[35] In this the sample was first treated with a cation exchange resin to remove interfering cations. The ability of the sample to complex zinc in the presence of Zincon reagent was then determined by measuring the absorbance of a zinc–Zincon reagent solution at 620 nm. The decrease in absorbance was a measure of the amount of NTA present. Other complexing agents that will complex zinc in competition with Zincon would also be measured as apparent

NTA. A method for determining the absorbance change using the Auto Analyzer® was described. The detection limit was about 0.2 mg l^{-1} NTA.

In the method described by Kaiser[36] the cobalt complex of the sample was prepared by evaporating one litre of the water sample to dryness in the presence of cobalt chloride. The absorbance of the solution was measured from 450 to 600 nm and the NTA complex was selectively oxidized by boiling with hydrogen peroxide. The absorbance over the wavelength range was again measured, the difference being proportional to the NTA level. The method was shown to distinguish between EDTA and NTA. The detection limit was 10 µg l^{-1} NTA.

Electrochemical methods have used the measurement of the complexing capacity of the sample for copper with the copper ion selective electrode.[1a] This method has been described by Bouveng et al.[37] Polarographic methods have also been used and determinations of the concentrations of NTA complexes of a number of metals, such as bismuth, indium and copper, have been described. A polarographic method using the bismuth complex has been automated in a form suitable for routine analysis.[38] The detection limit is 10 µg l^{-1}.

The gas chromatographic techniques offer a very sensitive and specific way of determining NTA. The procedure generally followed is to extract the NTA from the water sample by absorption on ion exchange resin (which removes interfering metal ions), elute it and prepare a derivative for the gas chromatographic determination. Rudling[39] describes a method where the NTA is adsorbed on an anion exchange resin, it is eluted with hydrochloric acid and esterified with a solution of boron trifluoride in 2-chloroethanol. The NTA tri-(2-chloroethyl-ester) is determined using an electron capture detector. The column packing was 2% QF-1 on Varaport 30. The identity of the peak as the NTA ester was confirmed by g.c.–m.s. Chau and Fox[40] eluted the NTA from the anion exchange resin with formic acid and prepared the propyl ester of the NTA. This was chromatographed on 3% OV-1 on Chromosorb WHP using a flame ionization detector. Warren and Malec[41] concentrated the sample by freeze-drying. The residue was treated with n-butanol–hydrochloric acid to convert the NTA to the n-butyl ester. Other aminocarboxylic acids that might be present were converted either to the n-butyl ester or to the N-trifluoroacetyl n-butyl ester derivatives. These were determined by gas chromatography. The detection limit was about 25 µg l^{-1} of the various aminocarboxylic acids. Aue et al.[42] describe a method for the determination of NTA and citric acid in water and sewage effluent. The sample was treated with anion exchange resin to separate the NTA and citric acid, and they were then eluted with formic acid. The n-butyl esters were then prepared by treatment with butanol–hydrochloric acid. The esters were dissolved in acetone for injection to the gas chromatograph. The separations were made on a column of Carbowax 20M on Celite 545, and a flame ionization detector was used. The detection limit was 1 µg l^{-1} of each of NTA and citric acid.

A summary of the analytical methods for NTA was made by Kaiser.[36] Water samples for NTA analysis are preserved either by acidification or by the addition of formaldehyde[41] and by low temperature storage.

ORGANOCHLORINE PESTICIDES AND POLYCHLORINATED BIPHENYLS

The organochlorine pesticides and polychlorinated biphenyls (PCBs) may be extracted from environmental samples and determined by chromatographic methods. The following procedure is described in *ASTM Standards*[43] for the determination of the organochlorine pesticides in water; their determination in other environmental samples is described below. The compounds that are determined with this method are: BHC, lindane, heptachlor, aldrin, heptachlor epoxide, dieldrin, endrin, Perthane®, DDE, DDD, DDT, methoxychlor, endosulfan, γ-chlordane and Sulfenone.

One litre of the water sample is extracted with 60 ml of a mixture of ethyl ether in hexane (15 + 85). The extraction is repeated, and then an extraction is made with 35 ml of hexane. The solvent extracts are dried, by passing them through anhydrous sodium sulfate, and combined. They are then evaporated to about 1 ml volume. An initial screening of this extract is made by gas chromatography to determine both the extent of any interferences which are present and the approximate concentration of the pesticides of interest. The extract is then diluted to about 10 ml with hexane and 'cleaned-up' by Florisil column adsorption chromatography.

The diluted extract is passed through a column of Florisil (about 15 g) with some sodium sulfate above it, where the organics are adsorbed. The organo-chloride pesticides are then eluted in two groups, the first group by passing 200 ml of a mixture of 6% ethyl ether in hexane through the column. This elutes the lindane, BHC, aldrin, heptachlor, DDE, DDD, DDT, Perthane®, heptachlor epoxide, methoxychlor, γ-chlordane, PCBs and endosulfan I. A second elution is made with 200 ml of 15% ethyl ether in hexane and this elutes the dieldrin, endrin, endosulfan II and some lindane.

These extracts are evaporated to a volume of about 1 ml in a Danish-Kuderna evaporator. The concentrated solutions are then examined by gas chromatography using an electron capture detector. The less sensitive microcoulometric detector may be used, but in this case a larger water sample is used. The columns that are used in the gas chromatograph are 3–6 mm i.d. about 1.8 m long. The column packings that are used are: the less polar 5% OV-17 on Gas Chrom Q, and the more polar 5% QF-1 plus 3% DC-200 on Gas Chrom Q. (As discussed below, it is desirable to use columns of different polarity in order to confirm the identification of the peaks observed.) A typical chromatogram of a mixture of some of the standards used is shown in Fig. 6.2.

An alternative procedure for cleaning up the extracts is to use thin layer chromatography (t.l.c.). Plates with a layer of Silica Gel G, 250 μm thick, are employed. The extracts are spotted on the plate, using up to 100 μl of the sample. Standards are applied to the plate at the same time. The plate is developed in a chamber using chloroform as the solvent. When the solvent front has migrated 10 cm the plate is removed from the chamber and air dried. The positions of the standards on the plate are then determined. This is done by covering the part

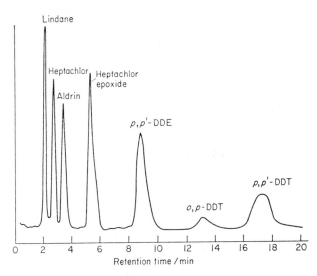

Fig. 6.2. Typical chromatogram of pesticide mixtures.

of the plate containing the sample, and spraying the area containing the standards with a solution of Rhodamine B in ethanol (0.1 g l^{-1}). The sprayed area is allowed to dry and then examined under a u.v. lamp. The areas containing the pesticide standards show up as dark areas in the fluorescent background. The part of the plate containing the sample is then divided into three sections, as shown in Fig. 6.3. These areas of silica gel are then removed from the plate and dissolved in solvent for gas chromatographic examination. The approximate position on the plate of some pesticide standards with their reported R_F values are also shown in Fig. 6.3.

Polychlorinated biphenyls (PCBs) may be separated from the other pesticides by adsorption chromatography, on a silicic acid column as described by Armour and Burke,[44] or on a column of activated charcoal as described by Berg et al.[45] Historically there have been some problems in quantification of PCB peaks; Berg et al.[45] prepared the bicyclohexyl and decachlorobiphenyl derivatives in order to quantify the PCBs. Webb and McCall[46] have described the problem of quantification using an electron capture detector and have suggested the use of three Arochlors (1242, 1254 and 1260) as standards for environmental analysis.

In the procedure described above the pesticides are separated from the water by solvent extraction. An alternative procedure is to extract the organic materials with a resin. In the method described by McNeil et al.[47] for the determination of chlorinated pesticides in potable waters, Amberlite XAD-2 resin was used. This is a low-polarity styrene–divinylbenzene copolymer; the adsorbed pesticides were recovered from the resin by extraction with hexane.

For samples other than water, such as vegetation, fish, soil and sediment, the samples are extracted with a solvent. Benzene, acetone–hexane mixture and

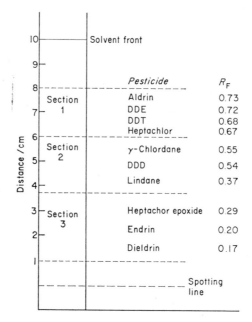

Fig. 6.3. Thin layer chromatographic separation of some chlorinated pesticides on silica gel G plate, chloroform solvent.

acetonitrile are some of the solvents that are used. The use of acetonitrile has the advantage that some cleaning up of the extracts takes place if the acetonitrile used for the extraction is diluted with water and this solution is then extracted with petroleum ether to recover the pesticides. For sampling air the particulates collected on a filter are extracted with a solvent. The volatile pesticides are collected by passing the filtered air through an impinger train containing ethylene glycol or hexylene glycol. The glycol solution is diluted with water and extracted with hexane to separate the pesticides. An alternative method is to use plugs of porous polyurethane foam to absorb the pesticide vapour from the air.[48]

In collecting any samples for pesticide analysis care must be taken to avoid contamination, particularly from some plastic materials. In general, the only materials that are suitable for handling samples are glass, aluminium foil and Teflon®. Hence water samples are collected in glass bottles with a Teflon® lining in the cap; similarly the other types of sample are collected in glass jars or are wrapped in aluminium foil.

With the gas chromatographic methods many of the chlorinated pesticides may be determined at the parts per trillion level.

VOLATILE ORGANOHALIDES

The volatile chlorinated hydrocarbons have as their source either the release of industrial solvents and aerosol propellants to the environment, their production by the chlorination of organic materials as, for examine, in water treatment or, in the case of carbon tetrachloride, by biochemical synthesis in the environment. The determination of these materials is mostly carried out by gas chromatography using an electron capture detector. In the case of water samples the choice between methods is that of the procedure used for separating the volatile organohalides from the water. In the method described by the Analysis Working Group of the Bureau International Technique des Solvents Chlorés (BIT-SC),[49] a 200-ml water sample was extracted with 10 ml n-pentane. The n-pentane was dried over sodium sulfate and a 1-μl portion was injected into the gas chromatograph. The column packing was 7% Dexsil-300 on diatomite and a temperature programme was used. With the method eleven solvents, including chloroform, trichloroethane, carbon tetrachloride, trichloroethylene and perchloroethylene, were determined in the range 0.01–10 μg l^{-1} water.

An alternative to solvent extraction for the separation of the materials is volatilization. In the method described by Bellar and Lichtenberg[50] the sample was purged with a stream of nitrogen. The purge gas passed through a trap containing an adsorbent (Chromosorb 103 or Tenox GC) which adsorbed the organic vapours from the nitrogen. The trap containing the adsorbents was then heated rapidly to desorb the organics which were swept into the gas chromatograph. A number of halogenated materials in water could be determined by this technique down to levels of about 0.5 μg l^{-1}. When this volatilization/trapping system is combined with gas chromatography–mass spectrometry the whole range of volatile organic materials in water may be determined.

Rather than using an adsorbent to trap the organics the technique of headspace analysis can be used. In this technique the water sample is equilibrated with a small volume of gas in an enclosed vessel and this gas is then swept directly into the gas chromatograph. In the method described by Kaiser and Oliver[51] a 60-ml sample of water was placed in a separating funnel with an air space of about 2 ml. The funnel was placed upside down in a water bath at 70 °C for 30 min, after which time equilibration was complete. A 5-μl sample of the air in the funnel was then taken with a syringe through the stopcock and injected into the gas chromatograph. The column packing was 10% OV-1 on GasChrom Q, with an oven temperature of 50 °C. The method would detect 0.1 μg l^{-1} of such organohalides as $CHCl_3$, $CHBrCl_2$, $CHBr_2Cl$ and CCl_4.

Water samples for the determination of the volatile organics are best taken in a sample bottle fitted with a septum cap. The bottle is filled completely and the cap containing the septum is applied. When the sample reaches the laboratory an aliquot of the sample can be taken through the septum, thus avoiding the loss of any volatile material. Where the sample contains residual chlorine, this should be destroyed by the addition of sodium thiosulfate or sodium bisulfite when the sample is taken. This prevents the further formation of organohalides during the transport of the sample to the laboratory. The sample should be preserved by storing and transporting at 4 °C (i.e. in an ice chest).

A method for the determination of the organohalides in air, described by BIT-SC,[49] is similar to the method for water. The sample is taken in a 250-ml flask fitted with a septum holder containing a silicone rubber septum. For the analysis a 2-ml aliquot of the air sample is taken with a syringe through the septum and injected into the gas chromatograph. The method measures the chlorinated solvent as above in the $1–100$ ng l^{-1} air range of concentrations.

In measuring the concentration of CCl_3F and CCl_2F_2 (also SF_6) in the atmosphere DeBortoli and Pecchio[52] collected the air samples in 'saran' (vinylidine chloride) bags. A 3-ml sample was injected into the gas chromatograph which used a column packed with 5-Å molecular sieve. The detection limits were 1×10^{-11} volume fraction for CCl_2F_2 and 2×10^{-12} volume fraction for CCl_3F.

ORGANOPHOSPHORUS PESTICIDES

Organophosphorus pesticides can be determined by most of the chromatographic methods. In contrast to the chlorinated pesticides a clean-up procedure is generally not required for gas chromatographic analysis because the flame photometric detector (FPD) in the phosphorus mode is very specific for phosphorus. A comprehensive scheme for the extraction of the organophosphorus pesticides from river water and sewage effluents is described by Askew et al.[53] They extracted a one-litre sample of water successively with three 50-ml portions of chloroform. The chloroform extract was dried and evaporated to small volume. In many cases this could be used directly for gas chromatographic analysis but, if a clean-up procedure was required, a column of activated carbon was used, the pesticides being eluted with chloroform. For gas chromatography the concentrated chloroform extract was freed from chloroform by evaporation with acetone or ethanol, and an ethanol solution was made for injection. Three column packings were used: 2% Apiezon L + 0.2% Epikote 1001; 4% S.E. 30 + 0.4% Epikote 1001; 2% X.E. 60 + 0.2% Epikote 1001, all on Chromosorb G. Pesticides were determined down to the 0.1 μg l^{-1} level.

Separations were also made by thin layer chromatography using a silica gel coating 250 μm thick. The solvent systems used were hexane–acetone $(5 + 1)$, chloroform–acetone $(9 + 1)$ and chloroform–acetic acid $(9 + 1)$. The position of the spots on the plates was determined by using spray reagents containing ammonium molybdate and stannous chloride to generate molybdenum blue. The pesticides on the plate were hydrolysed to phosphate by heating to 180 °C. The method determined the pesticides down to 1 μg l^{-1} in the water sample.

Gel chromatography was also used to separate the pesticides into groups. The pesticides were adsorbed onto a column of Sephadex and eluted with ethanol. This divided the pesticides into three groups which were then examined by gas or thin layer chromatography.

With these techniques 41 organophosphorus pesticides were determined.

A procedure for determining 14 pesticides in water is described by Ripley *et al.*[54] The pesticides determined and their approximate detection limits are:

	$\mu g\,l^{-1}$		$\mu g\,l^{-1}$
Azinphosethyl	0.10	Imidan	0.10
Azinphosmethyl	0.10	Malathion	0.010
Carbophenothion	0.020	Methyl parathion	0.010
Crufomate	0.050	Methyl trithion	0.020
Diazinon	0.010	Parathion	0.010
Disulfoton	0.005	Phorate	0.005
Ethion	0.010	Ronnel	0.010

In this procedure one litre of water was extracted with 10 ml of benzene. To this extract was added 0.5 ml of toluene and it was evaporated to a volume of approximately 0.5 ml. Portions of this concentrated extract were used directly for the gas chromatographic analysis. The columns used in the gas chromatograph were: 11% OV-17 + QF-1 on Chromosorb Q; 3.6% OV-101 and 5.5% OV-210 on Chromosorb W; 3% OV-225 on Chromosorb W HP. The columns were operated isothermally at 200 °C. The OV-17 + QF-1 column was the only one which separated all the 14 pesticides, but the other two columns were used for a confirmation of the identity of the pesticides. Confirmation of the identity of some of the pesticides by derivatization may be made by hydrolysing the pesticides and forming the pentafluorogenzyl ethers of the phenolic hydrolysis products.[55] Three of the widely used pesticides, fenitrothion, phosphamidon and dimethoate, are not well separated by this gas chromatographic procedure. A method for determining these materials using solvent partition is reported by Ripley *et al.*[56] The water sample is extracted with hexane, which extracts fenitrothion and leaves the other two pesticides in the water. These are then extracted with chloroform and the two organic solutions are then separately concentrated and determined by gas chromatography.

Another way to detect pesticides is by their enzyme inhibiting activity. The organophosphorus pesticides are toxic because they inhibit the enzyme cholinesterase, which is an essential component of the animal nervous system. In *Standard Methods*[1] this effect is used as the basis for a method of determining the organophosphorus and the carbamate pesticides in water. The result is expressed as the 'cholinesterase-inhibiting substances'. In the test the cholinesterase is used to catalyse the hydrolysis of 3,3-dimethylbutyl acetate (DMBA) to 3,3-dimethylbutanol (DMB). The DMB produced is extracted with carbon disulfide and determined by gas chromatography using a flame ionization detector. The inhibiting effect on this hydrolysis from the pesticides in the water is determined by measuring the reduction in the amount of DMB produced when the pesticides from the water are present. A parathion solution is used to standardize the determination. The enzyme-inhibiting effect of the organophosphorus pesticides has been used to visualize the spots of pesticides separated by paper chromatography.[57]

POLYNUCLEAR AROMATIC HYDROCARBONS

Much of the analytical work on polynuclear aromatic hydrocarbons (PAHs) was done as a result of the interest in the possibly carcinogenic material in tobacco smoke. The later concern was for these materials in the particulate matter in air samples and in environmental samples generally. The compounds are separated by a variety of chromatographic procedures and can be detected by ultraviolet–visible spectrometry, fluorescence, a flame ionization detector or a mass spectrometer, depending on the chromatographic procedure used.

Lao et al.[58] examined the PAHs in air particulates by gas chromatography–mass spectrometry. The filters from the air sample were extracted with cyclohexane and the PAHs were separated from the aliphatic and heterocyclic compounds by column chromatography on silica gel (the 'Rosen' separation). Both packed and surface coated open tubular (SCOT) columns were used in the gas chromatograph. More than 70 major PAHs having from 2 to 7 rings were separated and identified from an air sample.

Fox and Staley[59] determined PAHs in air particulates using high-speed liquid chromatography (h.s.l.c.). The filter containing the particulates was extracted with benzene. This was evaporated to small volume for injection into the chromatograph. Fluorescence spectroscopy was used to examine the separated materials. Eleven PAHs were identified.

A system is described by Novotny et al.[60] for the fractionation, analytical separation and identification of PAHs in complex mixtures. The system is applicable to any mixtures but the paper made particular reference to the analysis of airborne particulate matter. The PAHs were extracted, subjected to several liquid partition steps and then fractionated on a lipophilic gel. The collected fractions were further separated by h.s.l.c. The material separated by h.s.l.c. was then chromatographed using a capillary column gas chromatograph. The eluants from the liquid chromatograph were examined by spectral methods such as ultraviolet, fluorescence emission and nuclear magnetic resonance spectroscopy.

Giger and Blumer[61] extracted the PAHs from sediments, fossil fuels and other environmental samples. The samples were Soxhlet-extracted with methanol for 24 hours, benzene was added and the extraction continued for a further 24 hours. The hydrocarbons were partitioned from the benzene–methanol extract into n-pentane. This was washed, dried and evaporated to small volume. After removal of any elemental sulfur extracted, the solution was subjected to gel permeation chromatography in an open column using Sephadex LH-20. The fractions from this were then column-chromatographed on alumina–silica gel, to separate the PAHs from the saturated hydrocarbons and most olefins. The PAHs were further purified by forming an adduct with 2,4,7-trinitro-9-fluorenone. The uncomplexed materials were washed from this adduct with n-pentane. The adduct was then split by percolation over a silica gel column. The PAHs recovered were examined by mass spectrometry and were further fractionated by chromatography on alumina. This resulted in seven fractions being obtained; these were:

1. polychlorinated biphenyls
2. phenanthrene
3. anthracene, pyrene, fluoranthene
4. chrysene, benzanthracene
5. benzopyrenes, perylene
6. benzoperylene, anthanthrene
7. coronene

Fractions 2 and 3 were rechromatographed on alumina and the fractions were then examined by visible–ultraviolet spectroscopy. Many PAHs were identified but the mixtures were so complex that a classification by molecular weight was all that was possible in some cases.

Grimmer and Bohnke[62] determined PAHs in meat, poultry, fish, yeast, etc. They extracted the homogenized samples with hexane and found that, for the protein-rich samples such as meat and fish, a saponification with potassium hydroxide had to be carried out before the extraction in order to recover all the PAHs. The extracts were concentrated by liquid–liquid extractions; methanol–water–cyclohexane and N,N-dimethylformamide–water–cyclohexane; and by column chromatography on Sephadex LH-20. A gas chromatographic separation was made using a column of 5% OV-1 on GasChrom Q using a flame ionization detector. All the samples contained about 100 PAHs. Only the major components were identified and determined. However, the chromatogram constituted a g.l.c. profile which recorded both the known and the unknown PAHs.

Onuska et al.[63] determined PAHs in shellfish by gas chromatography. Using a short, wall-coated glass capillary column with a flame ionization detector, separation and identification of a large number of PAHs was achieved. The detection limits were a few micrograms per kilogram of sample for many of the hydrocarbons. The clean-up procedure employed was Rosen's procedure using chromatography on a silica gel column.

The use of a nematic liquid crystal for the liquid phase was found by Jaini et al.[64] to offer an improved separation of the PAHs in gas chromatography.

A review of the determination of PAHs in waters is given by Harrison et al.[65] Because of their low solubility they occur at low levels in water, so that a high concentration factor is needed in any extraction process. Solvent extraction with benzene was used by Strosher and Hodgson[66] in the analysis of lake water.

TRIAZINE HERBICIDES

The triazines used as herbicides may be determined by paper, thin layer or gas chromatography. A procedure using thin layer chromatography was reported by Abbot et al.[67] for the determination of eight triazine herbicides. Water samples were brought to pH 9 by the addition of ammonia and extracted with dichloromethane. The herbicides were extracted from soil samples by shaking the sample with ethyl ether in the presence of ammonia. This ethyl ether extract was dried and reduced in volume to 1 ml. This was added to 200 ml of 0.1 N hydrochloric acid and the aqueous solution was then extracted with ethyl ether. The ether extracts were discarded, the aqueous solution was made alkaline with ammonia and extracted with dichloromethane. The dichloromethane extracts from the soil and water samples were evaporated to dryness and resolubilized in 50 μl of hexane. Aliquots were spotted onto a silica gel plate and developed in a chloroform–acetone (9 + 1) mixture.

In the gas chromatographic determination, use is made of the alkali flame detector (AFD) in the nitrogen mode. McKone et al.[68] describe a procedure for the separation and quantification of atrazine, ametryne and terbutryne by gas chromatography using this detector. In the same paper two other methods of determination were compared, a polarographic and a spectrophotometric. In the polarographic determination the residue from the dichloromethane extraction was dissolved in methanol and polarographed in a 50% methanol solution; 0.01 N in sulfuric acid. Two peaks were obtained, one for atrazine (at −1.05 V relative to the mercury pool) and one for ametryne plus terbutryne (at −1.45 V). The spectrophotometric method was non-specific; the absorbance of the hydroxytriazine produced by the hydrolysis of all three herbicides was determined. The detection limits found for the three methods were: gas chromatography, $1 \, \mu g \, l^{-1}$ for each of the three herbicides; polarography, $5 \, \mu g \, l^{-1}$ of terbutryne and ametryne and $10 \, \mu g \, l^{-1}$ of atrazine; spectrophotometry, $10 \, \mu g \, l^{-1}$ of each herbicide.

CARBAMATES

Carbamates and phenyl urea pesticide residues in natural waters were determined by El-Dib[69] using thin layer chromatography. The water samples were acidified and then extracted with chloroform or dichloromethane. The extracts were spotted onto a plate coated with silica gel. Fifteen compounds were investigated. Six different solvent systems were used, the choice depending on the compounds to be separated. The plates were heated in an oven to hydrolyse the pesticides and were visualized with spray reagents. For carbamates and phenyl urea, which hydrolyse to yield phenols or heterocyclic enols, a spray reagent of p-dimethylaminobenzaldehyde was used to produce coloured spots. The phenylcarbamates and related ureas, which hydrolyse to form substituted anilines, were treated with sodium nitrite to diazotize them and then coupled with 1-naphthol to form the intensely coloured azo-dyes. For an identification of the spots they were removed from the plate, extracted with chloroform and examined by ultraviolet spectroscopy.

Cohen and Wheals[70] determined substituted urea and carbamate herbicides by a combination of thin layer and gas chromatography. Water samples were extracted with chloroform and the extract was dried and reduced in volume for spotting onto the thin layer plate. In soils and plant materials the procedure could not separate the herbicides from other materials that were co-extracted. With these samples the herbicides were hydrolysed to give the parent amine. Hence the determination would not distinguish between those herbicides which hydrolysed to give the same amine. The soil and plant materials were extracted with acetone, the acetone layer was diluted with sodium sulfate solution and extracted with chloroform. The chloroform was then reduced in volume and the hydrolysis carried out by refluxing with acetic–hydrochloric acid for 2 hours. The hydrolysis mix was washed with chloroform then made alkaline with sodium hydroxide and the amines extracted with chloroform. The chloroform was reduced in volume for spotting onto the plate. The extracted herbicides or amines were spotted onto silica gel-coated plates and separated using either chloroform or hexane–acetone (5 + 1). The 2,4-dinitrophenyl derivatives were then prepared *in situ* by spraying the plates with 5% hydrochloric acid and a solution of 1-fluoro-2,4-dinitrobenzene solution (4% in acetone). The plate was covered with a clean glass plate and heated in an oven at 190 °C for 40 minutes. The areas that contained the derivatives were then scraped off the plate and the derivatives were eluted from the silica gel with acetone. The acetone solutions were concentrated and were injected into a gas chromatograph equipped with an electron capture detector. The derivatives, being aromatic nitro compounds, had high electron-capturing properties. In this way the materials that were not separated by thin layer chromatography could be resolved by the gas chromatograph and vice versa. The combined thin layer–gas chromatograph procedure also offered a good way to quantify the materials. The herbicides determined were: Barban, Chlorbufam, Chlorpropham, Diuron, Fenuron, Linuron, Metobromuron, Monolinuron, Monuron and Propham. The detection limits were 1 µg l^{-1} in water samples and 20–50 µg kg^{-1} in soil and plant materials.

A procedure was described by Lau and Marxmiller[71] for the determination of Landrin® and related carbamates in corn tissue. The plant material was extracted with acetonitrile. This extract was washed with hexane and diluted with water. Hexane–ethyl ether (3 + 1) was used to extract the carbamates and this extract was then dried and concentrated. A clean-up was made using a column of alumina and activated carbon. The eluant from the column was evaporated to near-dryness and dissolved in ethyl acetate. Trifluoroacetic anhydride was then added and allowed to react overnight at room temperature. The derivatives were extracted and injected into the gas chromatograph. A column of 3% OV-17 or 2% Reoplex 400 on GasChrom Q was used with an electron capture detector. The detection limit was 20 µg kg^{-1}.

CHLORINATED PHENOXY HERBICIDES

Chlorinated phenoxy herbicides such as 2,4-D, 2,4,5-T and Silvex may be determined by gas chromatography. A method for their determination in water is described in *Standard Methods*[1] and in the ASTM Standards.[43] The water sample is acidified to pH 2 with sulfuric acid and extracted three times with ethyl ether. The ether extract is then boiled with potassium hydroxide solution to hydrolyse any esters, acids, etc., to the potassium salt of the phenoxy acid. The solution is acidified and the free acid is extracted with ether. The acids are too polar for convenient separation by gas chromatography and the methyl esters are prepared. The esterification is carried out by heating the extract with a solution of boron trifluoride in methanol (14% BF_3). The methyl esters are then cleaned up by column chromatography on florisil. The eluted material is concentrated and injected into the gas chromatograph. The column packing is 1.5% OV-17 plus 1.95% QF-1 on GasChrom Q, and an electron capture detector is used. The detection limits are 10 ng l^{-1} of 2,4-D, 2 ng l^{-1} of Silvex and 2 ng l^{-1} of 2,4,5-T.

These herbicides can also be determined by thin layer chromatography. In the method described by Ellerker et al.,[72] the sample was extracted with ether. This extract was evaporated to dryness and the residue dissolved in ethyl acetate. The separations were made on plates coated with a 60:40 mixture of Kieselguhr G–Silica Gel G. The solvent system was a mixture of liquid paraffin, benzene, glacial acetic acid and cyclohexane (1:3:2:14). After the separation the plates were dried, heated at 120 °C for 10 min, sprayed with 0.5% alcoholic silver nitrate and reheated at 120 °C. The separated herbicides appeared as black spots. The detection limit was 10 mg l^{-1} of the herbicides.

CHROMATOGRAPHY

In the determination of trace materials present in an environmental sample, particularly with regard to the organic compounds, there are four processes involved. These are:

(a) separation of the material from the matrix and from other trace materials present in the sample
(b) detection of the separated material
(c) quantification of the amount of material separated and the relation of this to the amount of material in the original sample
(d) identification of the material.

Chromatographic separations

Chromatographic processes offer a very effective way of carrying out separations. The term 'chromatography' (i.e. colour-writing) was first used by Mikhail Tswett in 1906. He passed an extract of plant pigments through a column packed with calcium carbonate and these were separated into a series of coloured zones on the column. Chromatography now refers to a group of processes in which a separation is effected by the distribution of a sample between two immiscible phases, one of these phases being stationary. (This last is to distinguish chromatography from such processes as solvent extraction.) In practice the mobile phase is a liquid or gas; the stationary phase is a liquid or solid. When the stationary phase is a liquid it is generally supported by a solid material or by the walls of the tube itself. The combinations of phases plus the different phenomena responsible for the separation lead to a classification of the chromatographic systems, some of which are shown in Table 6.1.

TABLE 6.1
Classification of chromatographic systems

Mobile Phase	Stationary Phase	Phase Configuration	Mechanism of Separation	System Type
Gas	Liquid	Column	Partition	Gas–liquid (g.l.c.)
Gas	Solid	Column	Adsorption	Gas–solid (g.s.c.)
Liquid	Liquid	Column	Partition	Liquid–liquid (l.l.c.)
Liquid	Solid[a]	Column	Adsorption	Liquid–solid (l.s.c.)
Liquid	Solid[a]	Thin layer	Adsorption	Thin layer (t.l.c.)
Liquid	Liquid[b]	Thin layer	Partition	Thin layer (t.l.c.)
Liquid	Paper[a]	Sheet	Adsorption	Paper chromatography
Liquid	Solid	Column	Molecular size	Gel permeation chromatography

[a] May be ion exchange material; the system is then ion exchange chromatography.
[b] Adsorbed on solid matrix.

Regardless of the mechanism by which the separation takes place the process of all the chromatographic separations is the same. This may perhaps best be understood by looking at a very simple model. Suppose we regard the solid phase as being divided into a large number of small segments. Into the first of these segments we introduce a volume of mobile phase containing an amount Q of the material that is to be separated. There will be an equilibrium set up in which the material is divided between the stationary phase and the mobile phase. If the apparent distribution coefficient of the dividing process is k then there will be $(k/(k + 1))Q$ in the stationary phase and $(1/(k + 1))Q$ in the mobile phase. If an equal amount of 'clean' mobile phase is now introduced into the front of the stationary phase it will displace the portion of the mobile phase containing the $(1/(k + 1))Q$ of material into the second segment of the stationary phase. The first segment of the stationary phase, which contains $(k/(k + 1))Q$ of material, will then be filled with mobile phase that initially contains none of the material to be separated. The mobile phases and stationary phases will then re-establish equilibria so that in the first and second segments the distribution will be as in line 2 of Fig. 6.4. A further displacement of the mobile phase will result in the amounts in the first three segments being rearranged as shown in line 3 of Fig. 6.4. As more mobile phase is added incrementally to the front of the stationary phase the process continues. In each segment the amount of material in the liquid phase is transferred to the succeeding segment and a new

	1st segment	2nd segment	3rd segment	
Quantity in stationary phase	$\left(\dfrac{k}{k+1}\right)Q$			
Quantity in mobile phase	$\left(\dfrac{1}{k+1}\right)Q$			1
Quantity in segment	Q			
Quantity in stationary phase	$\left[\dfrac{k^2}{(k+1)^2}\right]Q$	$\left[\dfrac{k}{(k+1)^2}\right]Q$		
Quantity in mobile phase	$\left[\dfrac{k}{(k+1)^2}\right]Q$	$\left[\dfrac{1}{(k+1)^2}\right]Q$		2
Quantity in segment	$\left(\dfrac{k}{k+1}\right)Q$	$\left(\dfrac{1}{k+1}\right)Q$		
Quantity in stationary phase	$\left[\dfrac{k^3}{(k+1)^3}\right]Q$	$\left[\dfrac{2k^2}{(k+1)^3}\right]Q$	$\left[\dfrac{k}{(k+1)^3}\right]Q$	
Quantity in mobile phase	$\left[\dfrac{k^2}{(k+1)^3}\right]Q$	$\dfrac{2k}{(k+1)^3}$	$\left[\dfrac{1}{(k+1)^3}\right]Q$	3
Quantity in segment	$\left[\dfrac{k^2}{(k+1)^2}\right]Q$	$\left[\dfrac{2k}{(k+1)^2}\right]Q$	$\left[\dfrac{1}{(k+1)^2}\right]Q$	

Fig. 6.4. Distribution of material between segments on successive displacements.

equilibrium is established between the material in the stationary phase and the material coming in from the preceding segment. After n displacements the distribution of the quantity of material in the segments is an expansion of the binomial $[Q/(k+1)^n](k+1)^n$, i.e. in the pth segment the amount of material is

$$^nC_p\left[\frac{k^{n-p}}{(k+1)^n}\right]Q$$

When n is large, distribution is the typical Gaussian curve shown in Fig. 6.5. Hence the material is progressively moved through the stationary phase, the rate of its movement being determined by k. Thus, with materials that have different 'k' values, the relative rates of movement through the system are different and they become separated, the separation being of the form shown in Fig. 6.5.

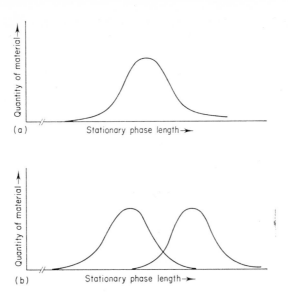

Fig. 6.5. Distribution of material as it is transported through the stationary phase. (a) Distribution of material after many 'displacements'. (b) Separation of two materials having different 'k's.

This type of model of a chromatographic separation was used by Martin and Synge[73] in considering gas chromatography by analogy with the separation achieved in a distillation column. The segments of the column considered were of such length that within each one there was complete equilibration of the vapour between the gas and the stationary phase. These zones, called 'Theoretical Plates', were considered analogous to the Theoretical Plates in a distillation column, and the lengths of the segments were analogous to the 'Height Equivalent to the Theoretical Plate' (HETP) used in a packed distillation column. Some of the imperfections of the model are that the chromatographic process is a continuous rather than an incremental one; that there is in practice radial and longitudinal mixing in the mobile phase; and that equilibrium between the mobile and stationary phases is generally not reached. However, the efficiency of a chromatographic column in producing separations is usually expressed in terms of 'number of plates'. The mobile phase and the material being separated move in the same direction. Hence the separation efficiency with a certain number of plates is much less than that which would be obtained with the same number of plates in, for example, a distillation process where the condensed liquid and the vapour move in a counter current fashion. The advantage of the chromatographic process is that it is much easier to produce and use a column that is equivalent to many thousands of plates than it is to prepare and use a distillation column with hundreds of plates.

In the various chromatographic systems the separation is utilized in different ways. In 'normal' analytical gas–liquid and gas–solid chromatography the process is continued until the materials are eluted from the end of the column. The time that each material remains in the column, the retention time, is characteristic of the material and (ideally) each material appears sequentially in the effluent gas from the column. To improve the separation process the 'k's may be changed during the process by changing the temperature of the column with a temperature programme. In liquid–liquid and liquid–solid chromatography using a column the material is eluted from the system. In these cases the 'k's can be changed by using different solvents or solvent mixtures in the mobile phase. In thin layer and paper chromatography the separated materials remain in the stationary phase at the end of the chromatographic process and they may either be identified and quantified *in situ* or they may be physically removed for off-line analysis.

Gas chromatography

The choice of a chromatographic technique is governed as much by the detectors available as by the separations that can be obtained. The large majority of organic materials in environmental samples are analysed by gas chromatography. This is probably because of the very high sensitivity obtainable with the available detectors. For example, an electron capture detector (see below) has a sensitivity to organic halides of about 2×10^{-14} g s^{-1}. A large part of the early work on the determination of trace organics in environmental samples was directed towards the chlorinated pesticides such as DDT. The electron capture detector proved to be an extremely valuable tool in this work.

A typical gas chromatography system is shown in Fig. 6.6. It consists of a column inside an oven with provision for very precise and uniform temperature control. A carrier gas, the mobile phase, is passed through the column at a uniform flow rate and flows from the column to a detector system. Provision is made to add the sample to the front of the column; liquid extracts are generally injected from a syringe through a septum. A typical volume injected is a few microlitres. Materials which have been collected in air sampling by adsorption

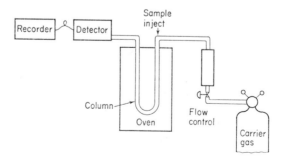

Fig. 6.6. Schematic of gas chromatography **system.**

onto a solid adsorbent are introduced by heating the adsorbent to release the material and sweeping it into the column. A typical system is shown in the description of the determination of the selenium alkyls (p. 55). The column may be a 2–6 mm i.d. column a few metres long, packed with the stationary phase; or it may be a capillary column about 0.2–0.6 mm i.d. many metres long, in which the stationary phase is a coating on the walls.

In a typical analysis, for example for chlorinated pesticides, an extract of the solid or liquid sample is made in a solvent. This separates the pesticides from the sample and provides a degree of preconcentration. Some preliminary 'clean-up' of the extract is then carried out, the exact procedure depending on whether the sample was of water, vegetation, fish, etc. The clean-up procedure is designed to remove other organic materials that may interfere in the gas chromatographic analysis. This clean-up may involve partitioning between various solvents, and usually a preliminary liquid chromatographic separation. This separation has traditionally been carried out by open column chromatography. The solution containing the extracted pesticides is passed through a short column of florisil (magnesium silicate), alumina or silica gel, where the organics are adsorbed. The desired organics are then obtained by elution with a solvent and collection of the appropriate fraction. More recently high-speed liquid chromatography (see below) has been adopted for this clean-up process. The concentrations that are achieved in this preliminary procedure are of the following orders: water samples, 1 litre to 1 ml; biological samples such as fish, 1 g of sample to 1 ml of extract; vegetation, 10 g of sample to 1 ml of extract.

Gas chromatography detectors

One of the earlier detectors used in gas chromatography was the thermal conductivity detector (TCD). This consists of two resistance wires in a Wheatstone bridge circuit. The two wires are heated electrically. One of them is placed in the carrier gas effluent from the chromatographic column and one is placed in a stream of 'clean' carrier gas. The resistance of the elements is a function of their temperature and this in turn depends on the thermal conductivity of the gas surrounding the elements. As materials of different thermal conductivity from the carrier gas are eluted from the column, the change in resistance is recorded by the bridge circuit. This detector is inexpensive and gives a response to almost every material that is analysed. However, it is not very sensitive (of the order of 1×10^{-10} g s^{-1}), and in environmental work has essentially been displaced by the various detectors described below.

An ionization detector consists of an ionization source and an electrode system that measures the ionization current. The effluent from the column is passed through the ionization chamber. A measure is made either of the ions produced by the sample material, as in a flame ionization detector, or of the reduction in the current by the electron-absorbing material in the sample, as in the electron capture detector. In the flame ionization detector (FID) the effluent is mixed with hydrogen and burnt in air or oxygen. Essentially all of the organic materials produce ions when burnt in the flame and the ionization current is measured. The detector is sensitive; it will detect 1×10^{-12} g s^{-1} of organic material. The alkali flame detector (AFD) is similar to the flame ionization detector but it runs with a higher fuel to oxidant ratio and on the burner jet is placed a pellet of an alkali metal salt, caesium bromide, rubidium sulfate or potassium chloride. This flame is then about 5000 times more sensitive to phosphorus compounds than it is to hydrocarbons, so that it can be used as a specific detector for organophosphorus compounds. By changing the gas flame it can also be made more sensitive to nitrogen compounds than hydrocarbons by a factor of about 50 to 1. In the phosphorus mode the sensitivity is about 5×10^{-14} g s^{-1} and about 5×10^{-12} g s^{-1} in the nitrogen mode.

In the electron capture detector (ECD) the carrier gas is ionized with ^{63}Ni or ^{3}H. The fall in ionization current when an electron-absorbing material enters the chamber is measured. Materials which are strongly electron absorbing are the organohalides and organometallics. The sensitivity of the ECD for organohalides is about 10^6 times its sensitivity to non-chlorinated organics.

A flame photometric detector (FPD) is similar in construction to the FID, but it measures the emission from the flame to determine sulfur (at 394 nm) and phosphorus (526 nm).

Another type of detector that is specific for chlorine and bromine compounds is the microcoulometric detector (MCD). In this the organohalide is burnt to form carbon dioxide and hydrochloric or hydrobromic acid. The halogen acid precipitates silver chloride or bromide in a cell containing silver acetate. An electrical imbalance is detected by the MCD and an amount of silver equal to that precipitated is generated electrochemically. The amount of electricity to do this is recorded. This detector is about 10 times less sensitive to chlorine than the ECD. The MCD can also be used in a different form to determine sulfur and phosphorus compounds.

A mass spectrometer can also be used as a detector; this is discussed below.

Quantitative analysis

The output from the gas chromatographic detector is normally recorded on a recorder chart, typically to give a trace as in Fig. 6.2 (p. 153). The output appears as a series of Gaussian peaks. If the sensitivity of the detector is the same for all the materials being reported, then from a mass balance it is possible to calculate the amounts of each material from the area under each peak and the 'total' area. Since this condition does not very often occur, the relative sensitivity of the detector to each material is determined experimentally. When these relative sensitivities are known it is possible to quantify the peaks by the use of internal standards. A known amount of a standard material that will give a peak separate from the 'sample' peaks is added to the sample extract. From a comparison of the areas of the 'standard' peak and the 'sample' peaks, the amounts of the materials in the sample can be calculated. Alternatively, and perhaps preferably, a known mixture of the materials being separated is injected and the peak areas recorded. This system of external standards allows the known mixture to be carried through the extraction procedure and any derivatization procedure so that the efficiencies of the processes can be monitored on a routine basis. The peak areas may be determined with an integrating recorder, or more commonly by an electronic integrator. Sometimes peaks overlap or one peak appears on the shoulder of another peak. There are mathematical treatments available to resolve these difficulties.

Identification and confirmation

When a material passes through a chromatograph, the retention volume with a given column is characteristic of the material. To relate the retention volume observed with a material to the retention volumes reported by other laboratories, the specific retention volume is calculated. This expresses the retention volume as a value at standard temperature and pressure per unit weight of stationary phase. This is often tedious to calculate and use is more often made of the relative retention times. When different materials are separated on a given column, although small differences in temperature, flow rate, pressure drop, etc., may change their absolute retention time, their relative retention times remain the same. Hence a retention time of an unknown may be determined by comparison with the observed retention time of a known material which gives a nearby peak. This is further refined by using the Kováts retention indices. In this technique, first proposed by Kováts, the retention times are related to the retention times of the n-paraffin series of different carbon numbers. By bracketing an unknown with two n-paraffins of appropriate carbon number, the retention time is very precisely determined as a Kováts retention index number. The use of this system in identification and in liquid phase characterization is discussed by Ettre.[74] Unfortunately gas chromatography by itself is a poor qualitative technique and the identification of unknown materials by retention time is only of use where the number of possible compounds is limited. This may be, for example, when a reaction has been carried out between two materials and it is desired to identify the few reaction products. In environmental samples the number of possible compounds is almost unlimited, and gas chromatography routinely is used only to quantify specific materials that are of interest. The retention indices, etc., are used to confirm that the retention times of the peaks being measured are the same as, for example, those of the standard pesticide materials. Even in such applications there are difficulties. In the chromatographic process, particularly with the 'traditional' packed column, it is quite possible for several materials to have the same retention time and appear as one peak. Thus, while the absence of a peak at a certain retention time is good evidence that a particular material is not there, the appearance of a peak at the retention time of a certain compound is not proof that that material is present. A quantitative determination based on the peak area assumes that all of the peak is due to the material of interest; if there are other materials under the peak, obviously the quantitative result is wrong. It is not uncommon to take an extract from a sample that gives perhaps 10 peaks with a packed column separation and to find that, when it is chromatographed using a higher-resolution capillary column, there are perhaps 30 peaks.

The problems of confirming the identification of materials found in environmental samples were briefly discussed in the section on metal alkyls, with particular reference to methylmercury. The techniques for the gas chromatographic confirmation of identity which are in common use are: using a variety of different separation conditions, making chemical derivatives of the material, and using a more specific detector.

In order to change the separation conditions the polarity of the liquid phase in the column may be changed. If possible the polarity is changed so that the materials in the sample elute in opposite orders in two columns. It is generally accepted that, if the peaks of the standards and of the samples retain the same retention times and the same quantitative relationship (within the experimental error) when chromatographed through three columns of different polarity, then the sample materials and the standards are the same material. An alternative is to use, for example, thin layer chromatography as one of the separation processes. Further confirmation can be obtained by removing the material that is separated as a spot on the thin layer plate and running it through the gas chromatograph. In this regard the use of liquid chromatography as a preliminary clean-up of the sample before gas chromatography (see above) offers the means of adding a further discrimination step into the analytical procedure.

In the derivatization process for confirming the identification, a portion of the sample extract is treated chemically to convert the material of interest to another material. The chromatograms run before and after the treatment confirm that the treatment material disappears and the new material is formed. A very common treatment is to use sodium hydroxide to dehydrochlorinate the chlorinated pesticides to form the corresponding olefin. Thus DDT when treated with sodium hydroxide will be converted to DDE.

Using detectors that have some specificity for the material being analysed provides some verification of its identity. If the material of interest contains the elements sulfur, nitrogen, phosphorus or chlorine, then a comparison of the results obtained with the different detectors described above can give an indication of whether these elements are present in the material giving the peak. The detector that gives much more information about the identity of the materials being separated is the mass spectrometer. The coupled gas chromatograph–mass spectrometer gives the analyst the opportunity to answer the question 'What is in the sample?', rather than just the question 'How much of this pesticide is present in the sample?'; the system is described below. If a sufficient amount of the material can be collected, the identity can be verified by such techniques as infrared absorption or nuclear magnetic resonance. In most cases, however, it is not practical to do this with environmental samples. This discussion on the identification-confirmation of materials is given in the section on gas chromatography because this is how the majority of pollutants in environmental samples are currently analysed. The problem exists, however, in all the forms of chromatography.

Gas chromatography–mass spectrometry (g.c.m.s.)

One of the most powerful techniques available for positively identifying specific organic materials in environmental samples is mass spectrometry. For this purpose it is most often used directly coupled to a gas chromatograph, and this g.c.m.s. is used both to confirm the presence and levels of specific materials in the gas chromatograph effluent and to identify unknown compounds. The mass spectrometer separates charged molecules and atoms by the mass to charge ratio. Its operation will not be discussed here, but it should be noted that it is an instrument that will respond to nanogram levels of material which makes it suitable for examining the small amounts of material eluting from a gas chromatograph. With an on-line data handling facility, the large amount of data produced when a material is examined can be sorted and presented very quickly so that such a g.c.m.s. system can be used on a routine basis. A typical print-out of a mass spectrum of phenanthrene is shown in Fig. 6.7.

The organic molecule is ionized and fragmented. The mass spectrometer records the amounts and mass numbers of the charged fragments and from this spectrum the original molecule can be identified. The mass spectrometer operates at high vacuum (at least 1×10^{-5} Torr) and the gas chromatograph operates at atmospheric pressure, so that there is need for an interface between the two. This interface may be in the form of a molecular separator where an enrichment of the large organic molecules relative to the carrier gas by means of their size or inertia takes place. The material entering the mass spectrometer system is ionized and fragmented. This can be done by a variety of methods such as electron impact, chemical ionization or field ionization.

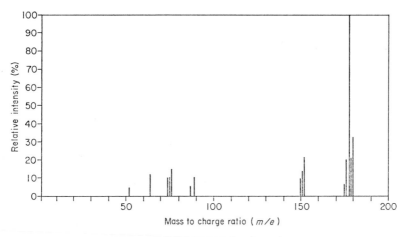

Fig. 6.7. Mass spectrum of phenanthrene.

In ionization by electron impact the molecules are bombarded with a stream of electrons. These electrons are emitted from a hot filament and are accelerated through the ionization chamber towards an anode. These electrons have high energies and tend to fragment the molecule rather than causing the production of the molecular ion.

In chemical ionization the sample flow is mixed with a large amount of a gas such as methane or isobutane, and the mixture is subjected to electron bombardment. Because of its high concentration relative to the sample, most of the electron interactions take place with the methane to form a plasma containing, for example, the CH_5^+ ion. This then reacts with the sample molecules to cause their ionization. This is a more 'gentle' treatment than electron impact so that a different spectrum of sample fragments is obtained, and there a higher proportion of molecular ions is formed.

Field ionization is also a technique that gives ionization at a relatively low energy. In this procedure the sample is passed through a very high positive electric field around a blade or wire held at a high positive voltage (i.e. 7–10 kV). The large majority of the ions formed are positive and the positive ion spectrum is the one usually examined. However, some negative ions are also formed and in some cases the less complex negative ion spectrum can be used.

The g.c.m.s. system can be used most simply to confirm the quantitative results of the 'routine' gas chromatographic analysis. The question in this case is whether all of the material in the peak in the chromatogram is the material that it is thought to be, or if there is some other material present. The sample extract is chromatographed again and the portion of the effluent that contains the 'peak' under test is introduced into the mass spectrometer. A comparison is made of the spectrum obtained from the sample with that obtained from the standard material. If any significant amount of materials other than that of the standard material are present in the sample 'peak' then this will be seen by the presence of extraneous fragments in the mass spectrum. However, this will not tell if an isomer of the standard material is present since this may give the same spectrum as the standard.

When the g.c.m.s. is used to determine the identity of unknown organic material in a sample, its mass spectrum is first compared with 'libraries' of mass spectra. There are a number of systems of classifying the mass spectrum and most of these now have the capability of being 'computer-searched' to find the ones which correspond most closely with the observed spectrum of the unknown material. If the identity of the unknown material cannot be established from the library search, then the structure may be determined from the information contained in the mass spectrum. The interpretation of mass spectra is described by McLafferty.[75] Briefly the steps he describes are as follows.

The elemental composition of the fragments is determined. This is done by considering the isotopic abundance of the elements. Many of the elements have isotopes that differ in atomic mass by one or two (or more) units. For example, carbon consists of ^{12}C and ^{13}C, the ratio of the two isotopes being 100:1.08.

The isotopes of the common elements and their relative abundances are shown in Table 6.2. The elements are designated as being 'A', or 'A + 1' or 'A + 2'

TABLE 6.2
Relative abundance (R.A.) of isotopes of some common elements

Element	Mass	R.A.	Mass	R.A.	Mass	R.A.	Type of Element
H	1	100	2	0.016	—		'A'
C	12	100	13	1.08	—		'A + 1'
N	14	100	15	0.36	—		'A + 1'
O	16	100	17	0.04	18	0.20	'A + 2'
Si	28	100	29	5.1	30	3.4	'A + 2'
P	31	100	—		—		'A'
S	32	100	33	0.80	34	4.4	'A + 2'
Cl	35	100	—		37	32.5	'A + 2'
Br	79	100	—		81	98.0	'A + 2'

types depending on whether their isotopes are not significant, have a mass number one unit greater, or have a mass number two units greater. Then if a fragment has a certain mass number and it contains an 'A + 2' element, the m/e spectrum will show another fragment two units higher, the relative intensities being the ratio of the abundances of the two isotopes. By looking first at the fragments that show this A + 2 pattern, and considering the relative intensities, it is possible to decide which of the 'A + 2' elements are likely to be in the fragment. A similar procedure is then gone through, with the 'A + 1' elements being considered. Then the balance of the mass of each fragment must be due to the 'A' elements and these can be assigned, remembering that the ratio of elements in a fragment must obey the laws of chemical bonding. Also all the fragments must come from the same original molecule.

From the valencies of the elements, it is possible to calculate the number of rings and double bonds in the fragments. In a molecule with the formula $C_xH_yN_zO_n$ the total number of rings and double bonds will be $x - \frac{1}{2}y + \frac{1}{2}z + 1$. (When the molecule is an ion the calculated value will end in a '$\frac{1}{2}$' and this is subtracted for the true value.)

The molecular ion M^+ is the 'fragment' that provides the most valuable information from the mass spectrum. This is the ion caused by the loss of an electron from the parent molecule. If this can be identified then the molecular weight of the unknown is determined and this places the limits on the possible elemental composition combinations that were calculated above. In the mass spectrum the fragment with the highest mass number may be the molecular ion, but it may not be so if the molecular ion is unstable. An indication whether the highest mass fragment is the molecular ion may be gained by applying the following criteria. The molecular ion is formed by the loss of an electron and the resulting ion is a radical, i.e. there is an unpaired electron remaining. Then if the calculation of 'rings plus double bonds' above is carried out for the suspected molecular ion, if it contains an unpaired electron the value of $x - \frac{1}{2}y + \frac{1}{2}z + 1$ will be a whole number. Another useful test, possible if the compound contains nitrogen, is the 'nitrogen' rule. For most elements that occur in organic compounds there is a correspondence between the valence of the element and the mass of its most abundant isotope, either the values are both even-numbered or odd-numbered. The exception is nitrogen which has an even mass number and one odd-numbered valence. This leads to the 'nitrogen rule' which states that: 'If a compound contains an even number of nitrogen atoms its molecular ion will be at an even mass number.' The other criterion to apply to a possible molecular ion is that it must be capable of giving the other smaller ions found in the spectrum by logical losses of neutral fragments and that the mechanisms by which the other fragments are postulated to arise must be reasonable ones.

An example of how these procedures may be applied can be seen in the following case of the identification of 5,6-dihydro-2-methyl-1,4-oxathiin-3-carboxanilide-4,4-dioxide which is a metabolite of Vitavax. In Fig. 6.8 is shown

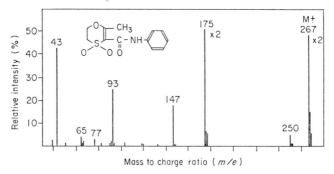

Fig. 6.8. Mass spectrum of 5,6-dihydro-2-methyl-1,4-oxathiin-3-carboxaniline-4,4-dioxide.

the formula of this material and the mass spectrum obtained with ionization by electron impact. From the group of peaks at 267 m/e the relative intensities are: 267, 97.5; 268, 14.3; 269, 5.8. For the calculation of elemental composition they may be transposed to:

m/e	Relative abundance	
267	100	P
268	14.7	P + 1
269	6.0	P + 2

Consider P + 1; the maximum number of carbon atoms possible is 13. Considering P + 2 there must be a sulfur or silicon atom plus another A + 2 atom (only one left is O). If silicon is present then the P + 1 abundance is reduced to

$(14.7 - 5.1) = 9.6$ which would leave room for only 8 carbon atoms. If one sulfur were present then the P + 2 abundance is reduced to $(6.0 - 4.4) = 1.6$, and the P + 1 abundance would be $(14.7 - 0.8) = 13.9$. This P + 1 abundance would leave room for 12 carbon atoms. If there are 12 carbon atoms these make a contribution of 0.8 to the P + 2 abundance. (This is because there is a chance of more than one ^{13}C atom being present in a group of 12 carbon atoms, and the contribution of the isotopic hydrogens.) This then leaves the P + 2 abundance as $(6.0 - 4.4 - 0.8) = 0.8$, which would leave room for 4 'O'. The P + 1 abundance is reduced to $(14.7 - 13.2 - 0.16 - 0.80) = 0.50$. The 'apparent molecular ion' is m/e 267 which, if it contains nitrogen, must contain an odd number of nitrogen atoms. For the residual P + 1 abundance above, and the mass left for completing a fragment of m/e 267, there can only be one nitrogen atom. Then the molecular ion formula is:

$$C_{12}H_{13}NSO_4 \qquad\qquad m/e\ 267$$
$$\text{Rings plus double bonds} \qquad 12 - \tfrac{1}{2}(13) + \tfrac{1}{2}(1) + 1 = 7.0$$

Hence m/e 267 is an ion containing an unpaired electron. m/e 267 is the highest mass observed and it can logically lose the neutral fragments OH and C_6H_6N to give the observed fragments m/e 250 and m/e 175 respectively. The examination of the P + 1 and P + 2 intensities of these other fragments is consistent with the projected formula. Hence the molecular ion does have the formula $C_{12}H_{13}NSO_4$.

Liquid chromatography

In its simplest (and original) form liquid chromatography consists of flowing a solution by gravity through an open column packed with the stationary phase. The separated materials may remain in the column and be recovered by dividing the column packing into segments or they may be sequentially eluted. In the modern more refined form of column liquid chromatography the stationary phase consists of particles of less than about 50 μm in diameter, packed in a column of 2–3 mm i.d. A flow rate of a few millilitres per minute is used and in order to obtain this flow rate pumping pressures up to about 5000 psig are required. This is referred to as High Speed Liquid Chromatography (h.s.l.c.) or High Pressure (Performance) Liquid Chromatography (h.p.l.c.). The stationary phase may be a solid adsorbent for liquid–solid chromatography, or an impervious solid, such as glass beads, coated with a liquid phase for liquid–liquid chromatography. The solvent used for the mobile phase determines the rate at which the materials elute from the column. Provision is usually made to change the composition of the solvent as the separation proceeds in order to accelerate, or retard, the elution of some of the components in the sample. A schematic of a typical h.s.l.c. system is shown in Fig. 6.9.

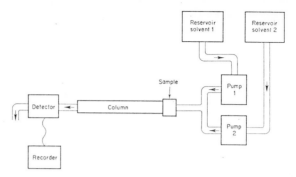

Fig. 6.9. Schematic of high speed liquid chromatography system equipped for gradient elution.

Detectors for liquid chromatography

In t.l.c. and paper chromatography the materials remain in the stationary phase and are measured *in situ*. In h.s.l.c. the sample materials elute with the mobile phase which is monitored continuously as in gas chromatography. The detectors which can be used are as follows.

An infrared absorption detector; this determines the absorbance at a wavelength where the components being separated strongly absorb radiation, comparing the absorbance to a known or an internal standard.

A refractive index detector; this senses the differential refractive index between solute and solvent.

A dielectric constant detector; this records the changes in dielectric constant of the solutions eluting from the column.

A titration-type detector where a titration such as acid-base polarographic or coulometric is made continuously.

A fluorescence detector where the fluorescence emission of the effluent is measured, using an excitation wavelength characteristic of the materials being separated.

An ultraviolet or visible spectrometer where the absorbance at specific wavelengths can be determined. The simplest ultraviolet absorbance instruments are those that measure at 254 nm wavelength using a low pressure mercury discharge lamp as the light source. For compounds which do not absorb strongly at 254 nm a spectrometer is used which can measure the absorbance over a range of wavelengths. This is probably the most widely used detector for h.s.l.c.; the detection limits under favourable conditions are 0.1–1 ng of material.

An electron capture detector is also used for h.s.l.c. The liquid effluent is fed to a travelling wire which passes through a heated chamber. The materials on the wire are vaporized and pass into an electron capture detector.

Quantification and identification

The calculation of the quantities of the materials separated is done in a similar manner to that used in gas chromatography, using either internal or external standards. The problems of positive identification are also the same and use is made of the same techniques as discussed above for assurance of correct identification of the separated materials.

Thin layer and paper chromatography

In thin layer chromatography (t.l.c.) the stationary phase is a thin layer of solid material applied to a rigid support such as a glass or plastic plate. The solid material particles may be coated with a liquid for liquid–liquid chromatography. The sample is applied as a spot or streak at one end of the plate and the plate is placed in a development chamber. This consists of a chamber with a pool of solvent in the bottom. The plate is placed on edge with its end in this pool of solvent and the liquid migrates up the plate by capillary action. The chamber is made reasonably air-tight and the atmosphere in the chamber is saturated with solvent to prevent evaporation of the solvent from the edges of the plate. The process is stopped when the various materials have been separated on the plate, i.e. before they have all migrated to the end of the plate. The spots or streaks are then visualized by fluorescence, by the application of a 'spray reagent' to the plate, or by thermal and chemical charring. It is possible in t.l.c. to obtain two-dimensional chromatography. After development with a solvent in one direction the plate may be turned through $90°$ and redeveloped with a solvent of different polarity. The extent of the separation of materials in t.l.c. is described by the R_F value. This is the distance travelled by the centre of the spot as a fraction of the distance travelled by the solvent front.

Paper chromatography is similar to thin layer chromatography, the difference being that the solid phase is a paper impregnated with solid adsorbent, liquid or ion exchange material. The same type of development is used as for t.l.c.

REFERENCES

1. *Standard Methods for the Examination of Water and Waste Water*, 14th Edn, APHA-AWWA-WPCF (1975).
1a. P. L. Bailey, *Analysis With Ion-selective Electrodes*, Heyden, London, 1976.
2. R. P. Cappelli, J. Lawrence and P. D. Goulden, *Measurement of Organics in Some Treated and Untreated Water Supplies of Southern Ontario*, FACCS Meeting, Philadelphia, U.S.A., November, 1976.
3. C. E. Van Hall, J. Safranko and V. A. Stenger, *Anal. Chem.* **35**, 315 (1963).
4. P. D. Goulden, *Water Res.* **10**, 487 (1976).
5. D. W. Menzel and R. F. Vaccaro, *Limnol. Oceanogr.* **9**, 138 (1964).
6. P. J. Williams, *Limnol. Oceanogr.* **14**, 292 (1969).
7. J. M. Baldwin and R. E. McAtee, *Michrochem. J.* **19**, 179 (1974).
8. M. Erhardt, *Deep Sea Res.* **16**, 393 (1969).
9. P. D. Goulden and P. Brooksbank, *Anal. Chem.* **47**, 1943 (1975).
10. Technicon Industrial Systems, Tarrytown, N.Y. (1976).
11. Y. Takahashi, R. T. Moore and R. J. Joyce, *Am. Lab.* **4**, 31 (1972).
12. C. E. Van Hall, D. Barth and V. A. Stenger, *Anal. Chem.* **37**, 769 (1965).

13. L. K. Wang, J. Y. Yang and M. H. Wang, *J. Am. Water Works Assoc.* **67**, 6 (1975).
14. N. Ciocan and D. F. Anghel, *Anal. Lett.* **9**, 705 (1976).
15. C. G. Taylor and J. Waters, *Analyst* **97**, 533 (1972).
16. P. T. Crisp, J. M. Eckert, N. A. Gibson, G. F. Kirkbright and T. S. West, *Anal. Chim. Acta* **87**, 97 (1976).
17. R. A. Greff, E. A. Setzkorn and W. D. Leslie, *J. Am. Oil Chem. Soc.* **42**, 180 (1965).
18. A. Nozawa, T. Ohnuma and T. Sekine, *Analyst* **101**, 543 (1976).
19. B. Baleux, *C.R. Acad. Sc. Paris* **274**, 1617 (1972).
20. L. Favretto and F. Tunis, *Analyst* **101**, 198 (1976).
21. R. Wickbold, *Tenside* **9**, 173 (1972).
22. R. M. Cassidy and C. M. Niro, *J. Chromatogr.* **126**, 787 (1976).
23. B. Stancher, F. Tunis and L. Favretto, *J. Chromatogr.* **131**, 309 (1977).
24. R. S. Tobin, F. I. Onuska, B. G. Brownlee, D. H. J. Anthony and M. E. Comba, *Water Res.* **10**, 529 (1976).
25. G. F. Longman, *Talanta* **22**, 621 (1975).
26. H. O. Friestad, D. C. Ott and F. A. Gunther, *Anal. Chem.* **41**, 1750 (1969).
27. B. K. Afghan, P. E. Belliveau, R. H. Larose and J. F. Ryan, *Anal. Chim. Acta* **71**, 355 (1974).
28. F. Drawert and G. Leupold, *Chromatographia* **9**, 605 (1976).
29. A. S. Y. Chau and J. A. Coburn, *J. Assoc. Off. Anal. Chem.* **57**, 389 (1974).
30. J. A. Coburn and A. S. Y. Chau, *J. Assoc. Off. Anal. Chem.* **59**, 862 (1976).
31. A. W. Wolkoff and R. H. Larose, *J. Chromatogr.* **99**, 731 (1974).
32. F. Dietz, J. Traud, P. Koppe and C. Rubelt, *Chromatographia* **9**, 380 (1976).
33. A. Wilander, H. Kvarnas and T. Lindell, *Water Res.* **8**, 1037 (1974).
34. S. H. Eberle, C. Hoesle, O. Hoyer and C. Kruckeberg, *Vom Wasser* **43**, 359 (1974).
35. J. E. Thompson and J. R. Duthie, *J. Water Pollut. Control Fed.* **40**, 306 (1968).
36. K. L. E. Kaiser, *Water Res.* **7**, 1465 (1973).
37. H. O. Bouveng, P. Solyom and J. Werner, *Vatten* **26**, 389 (1970).
38. B. K. Afghan, P. D. Goulden and J. F. Ryan, *Anal. Chem.* **44**, 354 (1972).
39. L. Rudling, *Water Res.* **5**, 831 (1971).
40. Y. Chau and M. E. Fox, *J. Chromatogr. Sci.* **9**, 271 (1971).
41. C. B. Warren and E. J. Malec, *J. Chromatogr.* **64**, 219 (1972).
42. W. A. Aue, C. R. Hastings, K. Gerhardt, J. O. Pierce, H. H. Hill and R. F. Moseman, *J. Chromatogr.* **72**, 259 (1972).
43. ASTM Standards, Part 31, American Society for Testing and Materials, Philadelphia, Pa. (1976).
44. J. A. Armour and J. A. Burke, *J. Assoc. Off. Anal. Chem.* **53**, 761 (1970).
45. O. W. Berg, P. L. Diosady and G. A. V. Rees, *Bull. Environ. Contam. Toxicol.* **7**, 338 (1972).
46. R. G. Webb and A. C. McCall, *J. Chromatogr. Sci.* **11**, 366 (1973).
47. E. E. McNeil, R. Otson, W. F. Miles and F. J. M. Rajabalee, *J. Chromatogr.* **132**, 277 (1977).
48. B. C. Turner and D. E. Glotfelty, *Anal. Chem.* **49**, 7 (1977).
49. Le Bureau International Technique des Solvants Chlorés, *Anal. Chim. Acta* **82**, 1 (1976).
50. T. A. Bellar and J. J. Lichtenberg, *J. Am. Water Works Assoc.* **66**, 739 (1974).
51. K. L. E. Kaiser and B. G. Oliver, *Anal. Chem.* **48**, 2207 (1976).
52. M. DeBortoli and E. Pecchio, *Atmos. Environ.* **10**, 921 (1976).
53. J. Askew, J. H. Ruzicka and B. B. Wheals, *Analyst* **94**, 275 (1969).
54. B. D. Ripley, R. J. Wilkinson and A. S. Y. Chau, *J. Assoc. Off. Anal. Chem.* **57**, 1033 (1974).
55. J. A. Coburn and A. S. Y. Chau, *J. Assoc. Off. Anal. Chem.* **57**, 1272 (1974).

56. B. D. Ripley, J. A. Hall and A. S. Y. Chau, *Environ. Lett.* **7**, 97 (1974).
57. M. E. Getz and S. J. Friedman, *J. Assoc. Off. Anal. Chem.* **46**, 707 (1963).
58. R. C. Lao, R. S. Thomas, H. Oja and L. Dubois, *Anal. Chem.* **45**, 908 (1973).
59. M. A. Fox and S. W. Staley, *Anal. Chem.* **48**, 992 (1976).
60. M. Novotny, M. L. Lee and K. D. Bartle, *J. Chromatogr. Sci.* **12**, 606 (1974).
61. W. Giger and M. Blumer, *Anal. Chem.* **46**, 1663 (1974).
62. G. Grimmer and H. Bohnke, *J. Assoc. Off. Anal. Chem.* **58**, 725 (1975).
63. F. I. Onuska, A. W. Wolkoff, M. E. Comba and R. H. Larose, *Anal. Lett.* **9**, 451 (1976).
64. G. M. Jaini, K. Johnston and W. L. Zielenski, *Anal. Chem.* **47**, 670 (1975).
65. R. M. Harrison, R. Perry and R. A. Wellings, *Water Res.* **9**, 331 (1975).
66. M. T. Strosher and G. W. Hodgson, *Water Quality Parameters* ASTM STP 573, American Society for Testing and Materials, 1975, p. 295.
67. D. C. Abbott, J. A. Bunting and J. Thomson, *Analyst* **90**, 356 (1965).
68. C. E. McKone, T. H. Byast and R. J. Hance, *Analyst* **97**, 653 (1972).
69. M. A. El-Dib, *J. Assoc. Off. Anal. Chem.* **53**, 756 (1970).
70. I. C. Cohen and B. B. Wheals, *J. Chromatogr.* **43**, 233 (1969).
71. S. C. Lau and R. L. Marxmiller, *J. Agr. Fd Chem.* **18**, 413 (1970).
72. R. Ellerker, H. Dee, F. Lax and K. Tucker, *Water Pollut. Contr.* **67**, 542 (1968).
73. A. J. P. Martin and R. L. M. Synge, *Biochem. J.* **35**, 1358 (1941).
74. L. S. Ettre, *Chromatographia* **7**, 39 (1974).
75. F. W. McLafferty, *Interpretation of Mass Spectra*, W. A. Benjamin, Reading, Mass., U.S.A., 1973.

CHAPTER 7

MICROORGANISMS

BACTERIA

Microorganisms, such as the different types of bacteria, are determined in environmental samples, often not because they are themselves of interest, but because they are known to originate in the digestive tract of higher animals, and thus are indicators of pollution arising from sewage or animal waste.

Samples are cultured in a medium in which bacteria will grow rapidly. After a certain time the effect of the multiplying bacteria can be measured or the growing colonies of bacteria can be counted. Distinction can be made between the different bacteria by selecting a culture medium and conditions which have some specificity for the growth of the bacteria of interest. Identification is made more certain by microscopic examination and testing of the bacterial colonies produced.

To determine the total bacteria in a water sample as a measure of its sanitary condition a 'standard plate count' is used. The procedure is described in *Standard Methods*.[1] The sample is mixed with a nutrient agar solution in a petri dish and incubated at 35 °C for 48 hours. After this time the colonies of bacteria that have grown on the plate are counted. The counting is either done visually with a microscope or with an automatic counting device. Dilutions of the sample covering several orders of magnitude are applied to different plates, so that a plate can be obtained where there are distinct individual colonies. The desired range of colonies for counting is 30–300 on a plate of approximately 65 cm² area.

The coliform bacteria are used as one indicator of pollution from animal wastes. These sources are the ones most likely to constitute a health hazard. The coliforms are defined in *Standard Methods*[1] as comprising all the anaerobic and facultative anaerobic gram-negative non-spore-forming rod-shaped bacteria that ferment lactose with gas formation within 48 h at 35 °C. In a sample being examined for coliforms it is desired both to establish their presence and also to determine their concentration. These tests are commonly carried out in fermentation tubes or by separating the bacteria on a membrane filter. The use of fermentation tubes has been standard practice for years but it is being replaced progressively by the more convenient membrane filter techniques as these are shown to be reliable.

In the procedure using fermentation tubes the sample is incubated with the appropriate nutrient solution at 35 °C for 48 hours. If gas is collected in the trap in the fermentation tubes, then a portion of the liquid from the tubes is planted in other tubes with another medium and reincubated. The production of gas in this second set of tubes is a further indication of the presence of coliform bacteria. This second fermentation is then used to plant bacteria on an agar plate. The colonies which grow on this plate are examined microscopically, first to determine their gram-stain characteristics (i.e. are they gram-negative) and second to determine that they are rod-shaped and non-spore-forming. From this examination the presence of coliform bacteria in the original sample may be confirmed.

The determination of the concentration, or bacterial density, in a sample is difficult because the bacteria are probably distributed unequally through the sample and it is almost impossible to take a small aliquot that is representative of the whole sample. The technique for determining the concentration using fermentation tubes is that of determining the 'most probable number' (MPN). This method is based on the fact that as a sample is diluted a point is reached where aliquots of the dilution would be expected to have a small chance of containing a single bacterium. If a sample is diluted progressively by factors of 10, at the point where there is one bacterium per millilitre, a 1-ml sample will have some chance that it will contain no bacteria. If a number of tubes, e.g. five, are planted with 1 ml portions of this dilution, the chances of all five tubes having no bacteria is small ($<1\%$). The chances of the sample 10 times more dilute than this yielding tubes with no bacteria is higher; the sample 10 times more concentrated obviously has a better chance of yielding all five tubes that contain bacteria. Hence the samples are diluted progressively by factors of 10 and portions of each dilution are planted in each of five tubes. The tubes are incubated and the production of gas monitored. A tube yields a positive result, i.e. gas formed, or a negative result, i.e. no gas formed. Using the results of sets of tubes covering three orders of magnitude that include the positive and negative results, the most probable number of bacteria in the original sample can be estimated. For samples containing low levels, plantings of 100, 10 and 1 ml of the sample are used.

A better indicator of pollution from sewage than total coliform bacteria are the faecal coliforms. These are the coliform bacteria that originate in the intestine of warm-blooded animals. Distinction is made between these faecal and other coliforms by planting the solution from the first fermentation described above in a different medium and incubating at 44.5 °C for 24 hours. If gas is produced in this second incubation it is indicative that the coliforms are of faecal origin. The faecal coliform density can be determined by the MPN technique.

In the membrane filtration technique for total coliforms an aliquot of the sample is filtered through an 0.45-μm membrane filter. (The size of aliquot is chosen so as to result in the growth of 50–200 coliform colonies.) The filter is then placed on an adsorbent pad that contains a nutrient medium and incubated at 35 °C for about 21 hours. In this incubation the bacteria grow colonies on the filter and these colonies can be counted. The nutrient medium is specific for coliforms and in addition the typical appearance of the coliform colonies is used to identify them. Distinction between faecal coliforms and the other coliform bacteria is made by incubation at 44.5 °C in an enriched lactose medium containing an indicator dye system. The faecal coliforms formed under these conditions absorb the dye and become blue in colour; any non-faecal coliform colonies are grey or cream-coloured.

One characteristic of the membrane filter technique that is useful is that it is possible to filter the samples in the field and transport the membrane filter to the laboratory. For this transportation the filter is placed on an absorbent pad saturated with a medium that keeps the bacteria alive but does not promote growth during the transportation time.

The definitions of 'coliforms' and 'faecal coliforms' used in the fermentation tube measurement and the membrane filtration method are not the same; hence comparison of the results by both methods in parallel is necessary to confirm that, in a particular type of sample, the results by the two techniques are equivalent.

The identification and quantification of the coliform bacteria by the MPN and the membrane filtration procedures are laborious and time consuming. A number of other ways of determining them have been proposed. In one of these[2] a lactose containing ^{14}C is used to prepare the nutrient medium in which the samples are incubated. Release of $^{14}CO_2$ as the coliforms metabolize the lactose is detected by trapping the $^{14}CO_2$ and measuring it in a scintillation spectrometer. Another method[3] uses an electrode system to sense the release of molecular hydrogen from a lactose medium. In another[4] the detection of the bacteriophage (i.e. the virus) that is capable of lysing the specific bacteria is used as a measure of the presence of the bacteria. These methods are reported to respond to the presence of coliforms in 6–8 h, compared to the 24-h period necessary in the membrane filtration test and the 48–96 h needed to carry out the tests in fermentation tubes. The system using the electrochemical detection of hydrogen is also proposed for use in a series of remote sampling stations.[5]

Other bacteria such as the faecal streptococci can be used as indicators of faecal pollution; the techniques used are similar to those described for the coliforms.

The use of bacteria as a means of detecting pollution from animal and human waste is dependent on the recovery of live bacteria and their culture. Under conditions where the bacteria do not survive in the receiving waters the indicator system breaks down. Such conditions arise, for example, when the sewage discharges are mixed with toxic industrial effluents or where a discharge from a primary sewage treatment is chlorinated. Although under such conditions the coliforms may not live, there is no guarantee that the other pathogenic organisms in the sewage discharge are also killed by the treatment. Hence an indicator of faecal waste other than the bacteria is desirable. It has been found that cholesterol is converted by bacteria in the lower intestine of mammals to coprostanol (5β-cholestan-3-ol). It is believed that this is the only route by which coprostanol occurs in nature and that the presence of coprostanol in a sample is indicative of contamination with mammalian faeces.[6] The analysis of coprostanol comprises a hexane extraction, a clean-up of this extract and determination by gas chromatography.[7] An alternative procedure uses a resin column rather than a hexane extraction to isolate the coprostanol from the water.[8]

VIRUSES

Viruses are detected and their concentration determined in a similar way to bacteria (i.e. by culturing the sample and observing the effect as the living viruses grow). However, since viruses can only multiply inside living cells, live organisms such as animals or tissue cultures are used as a medium for their growth. The concern in environmental samples is mainly for the possible health hazards of the enteric viruses. These are the viruses excreted by humans via the enteric tract. Tissue cultures are commonly used to examine samples for the presence of these viruses. These are cultures of living cells from, for example, a monkey kidney. Enteric viruses grow well on such cultures.

A suspension of the cells is put in a vessel, such as a dish or tube, with a nutrient medium. The cells attach themselves to the walls and multiply to form a monolayer on the inside of the vessel. When the sample containing viruses is introduced into the culture, the viruses infect some of the cells. The viruses multiply inside these cells and spread to neighbouring cells. This results in changes taking place inside the cells which then die. This destruction of the cells in the culture is known as the 'cytopathic effect' (CPE). The process of the virus spread from cell to cell can be slowed down by covering the monolayer of cells with a layer of agar. The CPE from one virus is then limited to a small area which looks like a hole in the monolayer of cells. These holes are called plaques. This effect can be used to determine the number of viruses in a sample by counting the plaques in the same way that the bacteria colonies are counted in the standard plate count. In this case the term 'plaque forming unit' (PFU) is given to the lowest concentration that forms one plaque. An alternative way of determining the concentration is to progressively dilute the sample and inoculate tissue culture tubes. These tubes are incubated and the dilution found that gives CPE in 50% of the tubes inoculated. This figure is known as the 'tissue culture infectious dose'—50% ($TCID_{50}$). Using the same technique as for bacteria the MPN can also be determined.

Positive identification of a particular virus is made by the use of antisera. This relies upon the fact that a specific antiserum will neutralize the effect of the virus against which it was prepared.

Many biological samples such as shellfish act to concentrate viruses so that a workable concentration may be obtained by homogenizing the tissue.[9] Viruses also adsorb onto particulate matter[10] so that soil and sediment samples are treated to desorb the viruses for analysis.[11] In the case of water samples, however, the virus concentration may be very low, perhaps 1 PFU per litre. In normal virology procedures the amount of sample used to inoculate a culture is a millilitre or less so that some form of concentration is required for the detection of viruses. In fact, the technology of virus determination in water is that of methods of concentrating the virus. One procedure is to use concentrated culture media so that most of the water going into the culture is that being tested. This allows assay of 10–20 times the sample volume that would otherwise be used.

A sampling method that concentrates the virus as it samples is the 'gauze pad method'. This is a modification of the 'Moore' swab method' for the isolation of bacteria in water. A gauze pad is immersed in the stream or lake being sampled, for periods of a few minutes up to several days. The viruses are adsorbed onto the pad and on removal of the pad can be desorbed from it. Further concentration of the desorbing liquid increases the sensitivity of the method. It has also been found that viruses are adsorbed on the matrix of a membrane filter when the water is pumped through it, even though the pore size of the filter is many more (e.g. 10–20) times larger than the virus. After a large volume of water is passed through the filter the viruses are eluted with, for example, a surfactant solution.[12,13] Separation of virus may also be made by ultracentrifugation. The larger particles such as bacteria are first removed by centrifuging at relatively low speed. Then the sample is centrifuged at about 60 000 g for an hour. These very high forces are necessary since viruses range in size from 20 to 200 nm. The water in a sample can be extracted by placing the sample in a dialysing bag immersed in a hydrophylic reagent such as polyethylene glycol. Viruses may also be concentrated by the phase separation method due to Albertsson.[14] This method uses the fact that if two polymers such as dextran sulfate and polyethylene glycol are added to water a two-phase system results. The particulate matter in the water distributes itself between the two phases, depending on its particle size and surface properties. Viruses will accumulate almost completely in one of the phases and hence can be concentrated. The procedure may be repeated to obtain a multi-stage process. Viruses will also adsorb onto particulate matter. A variety of precipitation reactions have been used to collect the viruses into a small volume of solid material. They can then be re-suspended in a small volume of water.

REFERENCES

1. *Standard Methods for the Examination of Water and Waste Water*, 14th Edn, APHA-AWWA-WPCF, 1975.

2. U. Bachrach and Z. Bachrach, *Appl. Microbiol.* **28**, 169 (1974).
3. J. R. Wilkins, G. E. Stoner and E. H. Boykin, *Appl. Microbiol.* **27**, 949 (1974).
4. R. P. Kenard and R. S. Valentine, *Appl. Microbiol.* **27**, 484 (1974).
5. J. R. Wilkins and E. H. Boykin, *J. Am. Water Works Assoc.* **68**, 257 (1976).
6. J. L. Murtaugh and R. L. Bunch, *J. Water Pollut. Control. Fed.* **39**, 404 (1967).
7. B. J. Dutka, A. S. Y. Chau and J. Coburn, *Water Res.* **8**, 1047 (1974).
8. C. K. Wun, R. W. Walker and W. Litsky, *Water Res.* **10**, 955 (1976).
9. M. D. Sobsey, C. Wallis and J. L. Melnick, *Appl. Microbiol.* **29**, 21 (1975).
10. G. Bitton, *Water Res.* **9**, 473 (1975).
11. S. deFlora, G. P. deRenzi and G. Badolati, *Appl. Microbiol.* **30**, 472 (1975).
12. H. A. Fields and T. G. Metcalf, *Water Res.* **9**, 357 (1975).
13. W. Jakubowski, W. F. Hill and N. A. Clarke, *Appl. Microbiol.* **30**, 58 (1975).
14. P. A. Albertsson, *Biochim. Biophys. Acta* **27**, 373 (1958).

CHAPTER 8

CONTINUOUS MONITORING

The analytical procedures described in the preceding chapters involve the sample being collected in the field and transported to the laboratory for analysis. An alternative approach to the determination of pollutant levels is to use equipment that samples continuously and carries out the analytical measurement at the sampling site. In this way the variations in the concentration of the pollutant are determined as a function of time. This continuous sampling and measurement is usually restricted to air and to water.

AIR MONITORING

The parameters which are most commonly monitored on a continuous basis are: oxidants, nitrogen oxide, sulfur dioxide, hydrocarbons, carbon monoxide, and particulates. These are perhaps the most important air pollutants in that these are the ones for which maximum permissible levels are defined in most air control regulations. The methods by which these are currently monitored are given in *ASTM Standards*[1] and are briefly described below. These are for the most part what may be described as 'first-generation' methods. The methods used in the laboratory are adapted so that they will operate in the field on a continuous basis. Many of the methods involve 'wet' chemistry, where the air sample is passed through an absorbent solution in which the pollutant of interest is separated. The concentration in the solution is then determined by the 'standard' chemistry using, for example, colorimetry or conductivity for the final measurement. In many cases the same type of equipment used in the automated continuous flow measurement in the laboratory is used in the continuous monitor. As in the laboratory methods, interferences are overcome by selective absorption and modifications to the 'measurement chemistry'.

In the 'second-generation' monitoring systems, which are now coming into general use, rather than using 'wet' chemistry, measurement is made directly on the air sample. Ideally the measurement made uses a property of the pollutant that is specific for that pollutant, so that manipulations to overcome interferences from other materials are unnecessary. The phenomenon which has proved to be most widely successful in these applications is that of chemiluminescence. This, and its application in air monitoring, is described below.

Oxidants

The continuous monitoring of oxidants in the atmosphere can be carried out using the reaction with potassium iodide that is described in Chapter 4. The oxidants in the air react with the potassium iodide to liberate iodine. The iodine is determined by measuring the absorbance at 360 nm with a recording colorimeter. A schematic diagram of the system used is shown in Fig. 8.1. The buffered

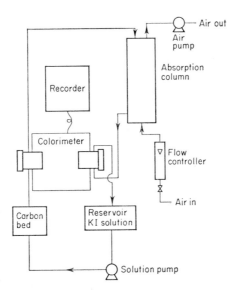

Fig. 8.1. Schematic of oxidant monitor.

potassium iodide solution is pumped from a reservoir through a bed of activated carbon to the top of the absorption column. The solution runs down the absorption column which contains a glass spiral to obtain good contact between the air sample and the liquid. The sampled air is drawn up the column at a known rate and is scrubbed in a counter-current fashion with the potassium iodide solution passing down the column. The liquid from the bottom of the column, which now contains the iodine produced by the oxidants in the air, passes to the absorption cell in the colorimeter, where the absorbance is continuously measured and recorded. The solution then passes back to the reservoir. The iodine that is in the reservoir as a result of the oxidation is removed from the solution by the activated carbon bed through which the solution is pumped on its way to the absorption column.

Ozone in the air may also be determined by measuring the nitrogen dioxide produced by the oxidation of nitric oxide with ozone. This method is described in the next section. The chemiluminescent method that is specific for ozone is described later.

Nitrogen oxides (and ozone)

Monitoring of the nitric oxide and nitrogen dioxide in the air is carried out using the Griess–Saltzmann reagent. This uses the chemistry described in Chapter 4 for the manual determination of nitrogen oxides in air. The nitrogen dioxide is absorbed in a solution of 0.5% sulfanilic acid and 50 ppm of N-(1-naphthyl)-ethylenediamine dihydrochloride in 5% acetic acid. The air to liquid ratio used is about 100 to 1. The air and the reagent solution are passed in a co-current fashion through an absorber; a glass tube wound into a spiral is commonly used for this absorber. The solution and gas coming from the absorber are separated. The liquid is passed through a delay coil, giving about a four-minute delay to allow the diazotization and coupling reactions to proceed to completion and the absorbance at 550 nm is continuously recorded and measured.

Nitric oxide in the air is measured by converting it to nitrogen dioxide. This is done by passing the sample through a bubbler containing potassium permanganate. The nitrogen dioxide is then measured in the sample before and

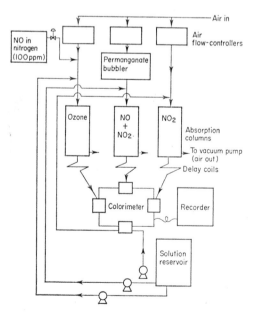

Fig. 8.2. Schematic of NO_2, NO and ozone monitor.

after the permanganate treatment, and the nitric oxide determined by difference. There is some loss of nitrogen oxides in the permanganate bubbler and a correction is made for this when the monitor is calibrated.

If there is any ozone in the air it will rapidly oxidize any nitric oxide to nitrogen dioxide, so that ozone and nitric oxide cannot exist together in the same sample. To determine ozone in a sample it is mixed with a stream of nitrogen containing about 100 ppm of nitric oxide, and the resulting nitrogen dioxide is determined as above. A correction is made for the nitrogen dioxide originally in the sample. A schematic diagram of this monitor is shown in Fig. 8.2.

Sulfur dioxide and sulfur gases

Sulfur dioxide in air is monitored by absorbing it from the air in a slightly acidified solution of hydrogen peroxide. The sulfur dioxide produces sulfuric acid in the solution and the conductivity is continuously measured to monitor this. This measurement is not specific for sulfur dioxide; it measures acidic gases such as hydrochloric acid as positive interferences, since these also increase the conductivity. Alkaline gases such as ammonia give a negative interference since the conductivity is reduced by the neutralization of the acid. However, unless the monitor is operated near a special source of interference, the conductivity measurement is a valid measure of the sulfur dioxide content of the ambient air.

An alternative detection method is to pass the air through an acidified solution containing bromine, and to determine the sulfur dioxide by a coulometric measurement. The bromine is produced by the electrolysis of potassium bromide. The bromine concentration is sensed by a redox electrode and, as the bromine reacts with the sulfur dioxide, more bromine is generated by the electrolysis reaction. The amount of current used to produce this bromine is recorded and is directly proportional to the amount of sulfur dioxide in the air being sampled. This method responds not only to sulfur dioxide, but also to other oxidizable sulfur compounds such as hydrogen sulfide, mercaptans, organic disulfides, etc. A schematic diagram of this technique is shown in Fig. 8.3. The West and Gaeke reaction[2] can also be used to monitor the sulfur dioxide in ambient air.

To distinguish between the sulfur-containing gases, such as sulfur dioxide, hydrogen sulfide and the mercaptans, etc., a technique involving a gas chromatographic separation has been used.[3] The air sample was automatically injected into the column in a gas chromatograph. The column would separate sulfur dioxide, hydrogen sulfide, methyl mercaptan and dimethyl sulfide. The detection used was the flame photometric detector in the sulfur mode. This method was the first successful one for separating the reactive sulfur gases; in earlier attempts there was trouble because of the reaction of the gases with the column and packing material. It was found that the only way to avoid this reaction was to use a column of Teflon® packed with powdered Teflon® coated with polyphenyl ether. As an air monitor this operates in a semi-continuous manner. An air sample is injected periodically to obtain the chromatogram of the sulfur compounds.

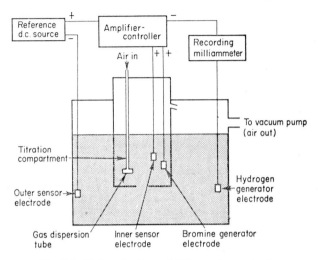

Fig. 8.3. Determination of SO$_2$ coulometrically.

Particulates

The concentration of particulate matter in air can be continuously monitored using optical methods. Where the concentration of the particulates is high, such as in smoke sources, the light absorption of the air sample can be measured directly. For ambient air samples the light absorption measurements are not sufficiently sensitive and light scattering techniques are used. In a typical monitor, such as that described in *ASTM Standards*,[1] the forward-scattered light is measured. In this technique the air sample is illuminated with light in a chamber; a phototube is at the opposite end of the chamber. A dark stop is installed in front of the light source so that the phototube is not directly illuminated, but the light that is scattered by the particulate matter through a small forward angle reaches the phototube. The advantage of the forward-scattering photometer, apart from its sensitivity, is that the forward scattering is not dependent on the refractive index and light absorbancy of the particles. A schematic drawing of such a monitor is shown in Fig. 8.4. The air sample is drawn through a coarse filter to remove particles greater than about 40 μm diameter. The air passes through the dark-field illuminated chamber where the forward-angle scatter is continuously measured and recorded. The air then passes through a weighed filter where the particulate matter is collected. Periodically this filter is removed and weighed. The weight of the particulate matter collected can be related to the optical transmission over the collection period. Then the mass concentration at any time can be determined from the optical transmission reading at that time. Calibration of the optical transmission reading can be made by passing an aerosol of a standard material through the illuminated chamber. The monitor will respond to particles of 0.05–40 μm in diameter.

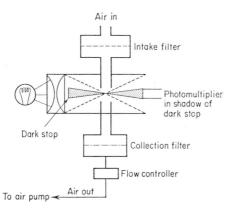

Fig. 8.4. Schematic of forward-scattered light photometer.

One type of this forward-scattering photometer can also be used to determine the size distribution of particles in a sample, since the forward scattering is a function of the particle diameter. In this application a small sensing zone is used so that the particles pass through it essentially one at a time, and the output from the photomultiplier is fed to a pulse height analyser. In this way the size distribution of the particles in an air sample may be rapidly determined.

An alternative means of measuring the particulates on a continuous basis is to filter the air through a moving filter tape. The amount of material collected on the tape may be determined by measuring the change in reflectance of the loaded tape relative to the unloaded tape. Another method of determining the mass collected on the tape is to place a β radiation source below the tape and to measure the absorption of the β radiation by the loaded tape.

Carbon monoxide

Carbon monoxide is monitored in ambient air by means of a non-dispersive infrared analyser. In this the air sample is passed through a long tubular cell. At one end of the cell is an infrared source, such as a heated element. At the other end of the cell is an infrared detector. Some selectivity in the wavelengths that are used can be obtained by the use of an optical filter in front of the detector. The amount of radiation that passes through the sample is compared with the amount passing through a reference cell that is filled with a gas of known composition. The most significant interference in the measurement is the absorption due to water vapour in the air. This may be overcome by chemically drying the air with a material such as silica gel, by controlling the humidity with a refrigerated trap, or by saturating both the air samples and the calibration gases with water vapour. The use of a narrow pass optical filter can also reduce the interference from water vapour. A schematic diagram of a typical monitoring system is shown in Fig. 8.5.

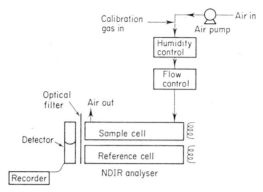

Fig. 8.5. Schematic of non-dispersive infrared analyser for monitoring carbon monoxide in air.

Absorption measurements are not very sensitive; the infrared techniques will only detect carbon monoxide at the parts per million level. To obtain a higher sensitivity a procedure is used that converts the carbon monoxide to methane. The methane is then determined by the very sensitive flame ionization detector. The carbon monoxide is first separated from the other components in the sample by gas chromatography. The carrier gas containing the carbon monoxide is mixed with hydrogen and passed over a heated nickel catalyst and then to the detector. This equipment will also measure methane and other hydrocarbons in the atmosphere, by passing the appropriate portion of the effluent from the gas chromatograph column to the flame ionization detector (see below). The equipment operates in a semi-continuous manner, like the sulfur gases separation and detection system above.

Hydrocarbons

Hydrocarbons in air are monitored with a flame ionization detector. If the sample is passed directly to the detector, the reading obtained is a response to the total number of carbon atoms in the sample. All carbon-containing materials give a response, not only the hydrocarbons. To obtain a measure of specific hydrocarbons (such as methane) the flame ionization detector is used as a detector after the materials in the air sample have been separated by gas chromatography. The equipment operates in a semi-continuous mode, like the other gas chromatographic monitors described above.

Chemiluminescence

Chemiluminescence, like fluorescence, involves the emission of light, as an atom or molecule falls from an excited state to a state of lower energy. It differs from fluorescence and other forms of luminescence in the way that the excited state originates. In the case of fluorescence the excitation is produced by the absorbance of photons from an incident radiation (see the description of atomic fluorescence in Chapter 2). In chemiluminescence the excited state is produced when an exothermic reaction takes place and some of the energy released remains in the reaction products as an electronic or vibrational excitation. The excited reaction products may lose their energy by radiation directly or they may transfer the energy by collision processes to other molecules in the system which in turn lose the energy. When part or all of these energy losses is by the emission of light, then the process is chemiluminescence.

Two types of chemiluminescence have been employed in air monitoring. The first is the reaction at ambient temperature of the air sample with a reactive material, most commonly a gas; the second is flame chemiluminescence. The reaction at ambient temperature is the more useful since it is much more specific, and it will be described first. In this process the air sample containing the pollutant 'P' is mixed with a reactive gas 'G' which reacts with P to produce an emitting species 'R*'.

$$P + G \rightarrow R^*$$

This then emits a photon of characteristic frequency.

$$R^* \rightarrow R + h\nu$$

The intensity of the light emitted is proportional to the concentration of R*, which is proportional to the product of the concentrations of P and G. The reactant gas G is generally in large excess so that the intensity of the light emitted is directly proportional to the concentration of the pollutant P. The light emitted is usually measured with a photomultiplier tube which may be fitted with an optical filter, to restrict the wavelength of light to which it will react to the wavelength characteristic of the emission from R*.

Chemiluminescence offers the same advantage as the other emission-type measurement in that by applying a high amplification to the signal from the photomultiplier it is possible to obtain a high sensitivity. The limits on the sensitivity are the dark current from the photomultiplier and the 'noise' in the amplifier circuits. These can be minimized by careful selection of the components. Hence the detection limits by this technique are much lower than those obtainable by the absorption spectroscopy techniques used for the direct measurement of such pollutants as carbon monoxide. In the determination of absorption, the signal is the difference between the two relatively large absorbances of the reference and the sample cell, each of which is subject to 'noise'.

The chemiluminescence system has a very high discrimination against positive interferences. In order for another material to give a spurious reading, it must first react with the reactant gas in an exothermic reaction to produce a chemiluminescence. In addition, the wavelength of the light emitted in this interfering chemiluminescence must be in the range transmitted by the filter to the photomultiplier. There are not many chemiluminescent reactions which occur at ambient temperature and under the above conditions the number of positive interferences in air monitoring applications are very few. The other type of interference which may occur is a negative interference from some other material quenching the excited species responsible for the light emission. The most common quenching agents encountered in air monitoring are molecular oxygen and nitrogen. Since these are at constant levels in atmospheric analysis, their quenching effect is a constant one.

That there are few chemiluminescent reactions at ambient temperature is an advantage in providing a specific determination. However, it is a disadvantage in that the applicability of the technique is limited. At the present time the technique is used for monitoring ozone, oxides of nitrogen and ammonia (by conversion to nitric oxide). The other possible applications appear to be the sulfur gases and carbon monoxide.

Ozone

Chemiluminescence most frequently occurs in reactions involving an energetic oxidation. Ozone is a strong oxidizing agent and there are a large number of materials, both inorganic and organic, which provide intense chemiluminescence when oxidized with ozone.

The first practical chemiluminescence detector for ozone was described in 1964 by Regener.[4] In this detector, Rhodamine B adsorbed on activated silica gel was reacted with the ozone in the air to produce the chemiluminescence. This is the

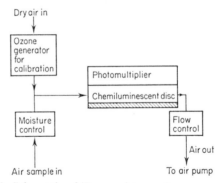

Fig. 8.6. Schematic of 'Regener' type of ozone monitor.

most sensitive method available for measuring ozone, with a detection limit well below one part per billion. The Rhodamine B-silica gel discs used in this monitor decrease in sensitivity as they are exposed to ozone, light or water vapour. Improvement in the life of the discs is made by preparing them with the incorporation of a silicone resin so that they are impervious to moisture. Even so, there is still a decrease in sensitivity with time, and it is necessary to calibrate the monitor frequently to correct for this change. A schematic diagram of this type of monitor is shown in Fig. 8.6.

Another ozone monitor uses the reaction between ethylene and ozone. This reaction was first described as a means of detecting ozone near nuclear accelerators by Nederbragt et al. in 1965.[5] At that time there was no further development of the technique. In 1970 Warren and Babcock[6] investigated the reaction as a means of providing a monitor for ozone in ambient air. They found that it had adequate sensitivity for ambient air measurements with a detection limit of a few parts per billion, and that the simplicity of the system makes for a low-cost, reliable monitor. This 'Nederbragt' monitor has since proved to be reliable and accurate in field service and is the type of monitor produced commercially by a number of companies. The ethylene–ozone reaction has been accepted in the U.S.A. as a reference method for the routine ozone measurements required by Federal air quality standards. The emission in this chemiluminescence is a broad band around 435 nm. The excited reaction product that gives the emission is believed to be an excited aldehyde linkage, e.g. formaldehyde, glyoxal. A schematic drawing of the 'Nederbragt' monitor is shown in Fig. 8.7.

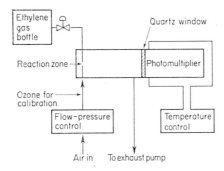

Fig. 8.7. Schematic of 'Nederbragt' type of ozone monitor.

There is also a chemiluminescent reaction between ozone and nitric oxide. This reaction is usually employed for measuring nitric oxide, although monitors, using this reaction, that will measure either nitric oxide or ozone have been described.[7]

Oxides of nitrogen

A number of chemiluminescent systems have been used for the monitoring of oxides of nitrogen. The reaction most commonly used is the reaction between ozone and nitric oxide.

$$NO + O_3 \rightarrow NO_2{}^* + O_2$$

$$NO_2{}^* \rightarrow NO_2 + h\nu \ (0.6\text{–}3 \ \mu m)$$

The kinetics and mechanism have been discussed by Clough and Thrush.[8] The emission is a modification of the air afterglow which is observed in electrical discharges through air or through nitrogen containing traces of oxygen. The emission is a continuum extending from 0.6 to 3 μm, and is red-shifted from 0.4–1.4 μm of the air afterglow. Air has a quenching effect on the chemiluminescence and the earlier monitors operated at a reduced pressure of about 1–5 Torr. This reduced pressure was produced by a mechanical vacuum pump. By modifying the design of the reaction chamber, to give a very small reaction zone, and increasing the sample flow, the quenching effect of the air can be reduced so that the reaction vessel can be operated at a pressure of 200 Torr or greater. This reduced pressure can be obtained conveniently with a diaphragm type of air pump, and most commercial monitors operate at this pressure. It has also been shown to be feasible to operate a monitor at atmospheric pressure. The detection limit of this NO–O_3 type of monitor is about 2–5 ppb NO. When the photomultiplier tube is cooled to $-25\,°C$ or lower, the monitor will detect levels of nitric oxide down to 1 ppb. The monitor response is linear over the range 0.001–10 000 ppm NO, i.e. it has a dynamic range of 1×10^7. An optical filter is used on the photomultiplier to cut off radiation of wavelength lower than 600 nm; this eliminates the interfering emissions from the reaction of ozone with other pollutants such as olefins.

In ambient air monitoring the oxide of nitrogen that is of interest, and the only one regulated by air quality standards, is nitrogen dioxide. In automotive emissions it is NO_x that is regulated. (NO_x is NO + NO_2.) In order to use the high sensitivity of the NO–O_3 monitor for ambient air, it is necessary to include a step by which the nitrogen dioxide in the air can be converted to nitric oxide. By making provision to by-pass this step, it is then possible to measure the total NO_x, the nitric oxide and (by difference) the nitrogen dioxide in the air sample.

The first work on the conversion of nitrogen dioxide to nitric oxide for monitoring by chemiluminescence was carried out by Sigsby et al.[9] They found that nitrogen dioxide could be converted quantitatively to nitric oxide by passing it through a stainless steel tube heated to a temperature of 600 °C or greater.

$$NO_2 \rightarrow NO + \tfrac{1}{2}O_2$$

Under these conditions any ammonia in the sample is oxidized to nitric oxide and hence interferes. This interference can be overcome by the use of acidic

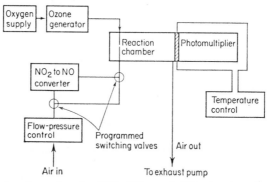

Fig. 8.8. Schematic of monitor for NO, NO_2 and NO_x by chemiluminescence with ozone.

scrubbers which remove the ammonia before it can be oxidized. However, the preferred way is to carry out the reduction at lower temperatures where the ammonia is not oxidized. A gold wool mesh heated to 240 °C has been used for this reduction, but the reduction appears to be accomplished by surface impurities and this type of convertor has an unpredictable lifetime. A variety of finely divided metal surfaces and metal surfaces impregnated with carbon have been employed. Carbon itself will also carry out the conversion and there are a number of commercial monitors that use carbon in various forms, such as graphite or carbon black. In Fig. 8.8 is shown a schematic of an $NO–O_3$ monitor that will distinguish between NO and NO_x by switching the air flow through or around the NO_2 reductor.

Another reaction that has been used for monitoring NO_x is that with atomic oxygen.

$$NO + O + M \rightarrow NO_2{}^* + M$$

$$NO_2{}^* \rightarrow NO_2 + hv$$

In this case the atomic oxygen first reacts with any nitrogen dioxide to form nitric oxide:

$$NO_2 + O \rightarrow NO + O_2$$

so that the technique is suitable for measuring the total NO_x content of the sample. The chemiluminescent reaction is the air afterglow reaction and the emission is a continuum from 0.4 to 1.4 μm. Prototypes of monitors using this reaction have been produced. The difficulties experienced with this technique have been in obtaining a source of atomic oxygen. Some of the monitors used a microwave or electrical discharge through oxygen or oxygen–argon mixtures. It was found, however, that there was always some nitrogen contamination in the gas and this produced an intense O + NO afterglow. This led to a fluctuating high background signal, which could be overcome by using a dual photometer system for background correction, but only at the cost of an expensive and complicated instrument. An alternative method for generating atomic oxygen, described by Black and Sigsby,[10] is to decompose ozone thermally. This appears to be a more flexible process for an NO_x monitor.

The use of photolysis to fragment molecules has been proposed as a way of extending the applicability of chemiluminescent techniques. In the case of nitrogen dioxide this will fragment to yield nitric oxide and atomic oxygen. The atomic oxygen can then be determined with nitric oxide:

$$NO_2 + hv \rightarrow NO + O$$

$$O + NO + M \rightarrow NO_2{}^* + M$$

$$NO_2{}^* \rightarrow NO_2 + hv$$

In this case, as in the NO_x monitor above, the chemiluminescent reaction is the air afterglow reaction with emission of a continuum from 0.4 to 1.4 μm. This photofragmentation technique enables nitrogen dioxide to be measured without a correction for any nitric oxide that might also be present.

Ammonia

As described above, ammonia can be oxidized to form nitric oxide. While this is a potential interference in the determination of nitrogen oxides, it can also be used as a means of monitoring ammonia. Using the O_3–NO monitor the air sample is pyrolysed to convert the ammonia to nitric oxide. The pyrolysis is repeated using a sample that has been passed through an acid scrubber to remove the ammonia. The difference is then a measure of the ammonia in the sample.

Flame chemiluminescence

In flame chemiluminescence the sample is passed into a flame, and the chemical reactions proceeding in the flame produce a luminescence of a characteristic wavelength. The chemistry that occurs in the flame is more complex than that of the ambient temperature reactions with reactive materials. The chemiluminescence usually results from reactions between atomic or molecular fragments which are produced from fragmentation in the flame of the molecules of the sample material. The flame chemiluminescence reactions are therefore much less specific than the ambient temperature reactions, since any material that contains the atoms or fragments that can be activated will produce a chemiluminescence. The most widely used application is in the detection of sulfur compounds. When sulfur compounds are burnt in a hydrogen-rich flame there is an intense blue chemiluminescence, consisting of a series of evenly spaced bands between 350 and 450 nm. This is from the diatomic sulfur S_2* formed by the recombination of sulfur atoms, and any sulfur-containing material will yield the chemiluminescence. To distinguish between different sulfur-containing pollutants in the air, it is necessary first to separate them (for example, by the chromatographic technique described above).

The flame photometric detector can also be used for determining nitrogen, phosphorus and boron in air samples. It is, of course, used as a means of obtaining an element-specific detector in gas chromatographic separations (see Chapter 6).

A review of chemiluminescent techniques for air pollution monitoring has been made by Hodgeson.[11]

Calibration

In all continuous-monitoring operations it is necessary to be sure that the results obtained on the read-out do represent the levels of the pollutant in the sample. A few methods are absolute ones, i.e. a measurement gives the absolute level of the pollutant of interest. However, most methods are comparative ones and it is necessary to analyse standard mixtures periodically in order to calibrate the equipment. Where the material analysed can be liquefied, permeation tubes are a convenient source of obtaining standard mixtures of gases. A permeation tube is a tube of Teflon® in which the liquid material is placed. The material permeates through the wall of the tube at a rate determined by the temperature. The permeation rates are generally of the order of a few milligrams per day. The permeation rate can be determined gravimetrically. Hence, if the tube is placed in a thermostated enclosure and a known volume of a gas such as nitrogen is passed over it, a gas mixture containing trace levels of the material can be obtained. Permeation tubes can be used to obtain standards of such materials as sulfur dioxide, nitrogen dioxide, ammonia, hydrocarbons, hydrofluoric acid and mercaptans.

Where the material cannot be obtained as a liquid it is necessary to make a standard mixture of gases that can be diluted with an inert gas to obtain the desired levels. In some cases, such as with ozone, it is not possible to obtain stable mixtures and the calibration is made by preparing an ozone mixture with an electric discharge under controlled conditions.

WATER MONITORING

A water sampling programme where the analyses are carried out in the laboratory enables the water from many locations to be examined for a large number of difficult-to-analyse pollutants. For regulatory purposes such a programme can also positively establish the presence of a pollutant at a certain level by giving the opportunity of confirmatory analyses by a number of techniques. However, changes that occur in the water from hour to hour or day to day can only be detected if the samples are taken very frequently. Alternatively, the water body can be continuously sampled and analysed to monitor on a continuous basis the changes in pollutant levels.

As in the case of air monitoring, continuous monitoring of water may be carried out by moving the laboratory equipment into the field. The continuous-flow automated methods are particularly suitable in this application and have been very successfully used both in shipboard and in shore-based monitoring programmes. Where the laboratory facilities can be duplicated beside the water body, any analysis that can be carried out in the laboratory can be performed on a continuous-monitoring basis. Duplication of the laboratory facilities requires that the equipment can be housed in a weatherproof, heated enclosure and that adequate services such as power are available. It also requires that access is possible on at least a daily basis for cleaning and servicing the equipment and replenishing such consumables as the reagent solutions. The requirement for continuous monitoring is for a system to bring the water to the continuously operating analytical system; this can be done with a conventional pumping system. The only 'unusual' requirement is that, for the determination of 'soluble' materials, a valid in-line filtration system is included in the delivery line.

Such installations are very expensive and are limited to those locations where they can be provided on a cost-benefit basis. Such situations are where it is desired to monitor the effluent from a factory on a continuous basis to ensure that any discharge regulations are met, and to be able to take corrective action quickly in the event of a spill. A number of 'automated wet chemistry' monitors are available commercially and they are most often used in this type of application.

For the monitoring of rivers and other water bodies, these types of monitor are generally too expensive to use, since they can only measure the pollutant levels at one particular place. The requirement in monitoring water away from the sources of pollution is for a low-cost piece of equipment that is capable of unattended operation in a remote location for periods of many weeks. In practice, this means that the only parameters that can be measured are those that can be determined by direct sensing devices. These devices at the present time comprise a conductivity meter, a thermometer, a turbidity meter and some of the ion-selective electrodes.[12] The parameters which can be measured by these electrodes are pH, dissolved oxygen, fluoride and chloride. These seven parameters have been measured successfully by remote unattended monitors over extended periods of time. Other electrodes such as those for ammonia, nitrate and carbon dioxide appear to be capable of being incorporated in such monitors, but they have not yet established a history of successful operation.

The problems in operating such monitors arise, in part, from the fact that the technology of measurement of water samples has for many years been developed in the laboratory. The laboratory provides a benign environment for instrumental measurements, with its steady ambient temperature, reasonable humidity levels, clean operating conditions and short operating cycles. To use the same technology in unattended operation in a remote location, where daily temperatures and humidities fluctuate widely, requires radical modification. In particular, there are severe problems of preventing the electrodes and equipment from being fouled by sediment and plant growth on their surfaces. These problems have been overcome by treating the exposed surfaces of the equipment with materials that inhibit algal growth, and by providing mechanical cleaning devices to scour the electrode and other measuring surfaces. Proper calibration also presents some problems since, although electronic calibration of the equipment is relatively simple, standard solutions for calibration of the whole measuring system suffer from the same problems of preservation that are encountered in the sampling of waters via bottles.

Continuous monitoring of parameters from the limited selection above is carried on as part of many programmes. The results give indication of changes in water quality rather than changes in the levels of specific pollutants. When the monitors show changes taking place, often the procedure followed is then to take samples by bottle to determine what these changes really mean. Hence, the continuous monitors are a supplement to a programme of sample collection and laboratory analysis, rather than a replacement for it.

REFERENCES

1. ASTM Standards, Part 26, American Society for Testing and Materials, Philadelphia, Pa., 1976.
2. F. P. Scaringelli, B. E. Saltzmann and S. A. Frey, *Anal. Chem.* **39**, 1709 (1967).
3. R. K. Stevens, J. D. Mulik, A. E. O'Keefe and K. J. Krost, *Anal. Chem.* **43**, 827 (1971).
4. V. H. Regener, *J. Geophys. Res.* **65**, 3975 (1960); **69**, 3795 (1964).
5. G. W. Nederbragt, A. Van der Horst and J. Van Duijn, *Nature (London)* **206**, 87 (1965).
6. G. J. Warren and G. Babcock, *Rev. Sci. Instr.* **41**, 280 (1970).
7. D. H. Stedman, E. E. Dalby, F. Stuhl and H. Niki, *J. Air Pollut. Control Assoc.* **22**, 260 (1972).
8. P. N. Clough and B. A. Thrush, *Trans. Faraday Soc.* **63**, 915 (1967).
9. J. E. Sigsby, F. M. Black, T. A. Bellar and D. L. Klosterman, *Environ. Sci. Technol.* **7**, 51 (1973).
10. F. M. Black and J. E. Sigsby, *Environ. Sci. Technol.* **8**, 149 (1974).
11. J. A. Hodgeson, *Toxicol. Environ. Chem. Rev.* **2**, 81 (1974).
12. P. L. Bailey, *Analysis with Ion-Selective Electrodes*, Heyden, London, 1976.

INDEX